TERTIARY LEVEL BIOLOGY

Physiological Strategies in Avian Biology

J.G. PHILLIPS, Ph.D., D.Sc., F.I. Biol., F.R.S.
Wolfson Professor
Wolfson Institute
University of Hull

P.J. BUTLER, Ph.D., F.I. Biol.
Professor of Comparative Physiology
Department of Zoology and Comparative Physiology
University of Birmingham

P.J. SHARP, Ph.D.
Senior Principal Scientific Officer
Agriculture and Food Research Council
Poultry Research Centre
Roslin
Edinburgh

Blackie

Glasgow and London
Distributed in the USA by
Chapman and Hall
New York

Blackie & Son Limited,
Bishopbriggs, Glasgow G64 2NZ

Furnival House, 14–18 High Holborn, London WC1V 6BX

Distributed in the USA by
Chapman and Hall
in association with Methuen, Inc.
29 West 35th Street, New York, NY 10001

© 1985 Blackie & Son Ltd
First published 1985

All rights reserved.
No part of this publication may be reproduced,
stored in a retrieval system, or transmitted,
in any form or by any means,
electronic, mechanical, recording or otherwise,
without prior permission of the Publishers.

British Library Cataloguing in Publication Data
Phillips, J.G. (John Guest)
Physiological strategies in avian biology——
(Tertiary level biology)
1. Birds——Physiology
I. Title   II. Butler, P.J.   III. Sharp, P.J.
IV. Series
598.2′1   QL698

ISBN 0–216–91780–8
ISBN 0–216–91784–0 Pbk

Library of Congress Cataloging in Publication Data
Phillips, J.G. (John Guest)
Physiological strategies in avian biology——(Tertiary level biology)
Bibliography: p.
Includes index.
1. Birds——Physiology
I. Butler, P.J.   II. Sharp, P.J.   III. Title
IV. Series
QL698.P43 1985   598.2′1   85–9689

ISBN 0–412–00921–8
ISBN 0–412–00931–5 (pbk.)

Photosetting by Thomson Press (India) Limited, New Delhi
Printed in Great Britain by Bell and Bain (Glasgow) Ltd.

Contents

Chapter 1.	GENERAL INTRODUCTION	1
Chapter 2.	LOCOMOTION	3
	2.1 Respiratory system	3
	2.2 Circulatory system	6
	2.3 Skeletal muscles	8
	2.4 Energetics of flight	8
	2.4.1 Power output	9
	2.4.2 Power input and efficiency	12
	2.5 Factors affecting power input (oxygen uptake)	16
	2.6 Physiology of forward flapping flight	18
	2.6.1 Respiratory adjustments	18
	2.6.2 Cardiovascular adjustments	22
	2.6.3 High altitude	23
	2.6.4 Future studies	24
	2.7 Hovering, gliding, bounding and undulating flight	25
	2.8 Take-off and landing	26
	2.9 Terrestrial and aquatic locomotion	27
	2.9.1 Walking/running	27
	2.9.2 Swimming	30
	2.9.3 Diving	33
Chapter 3.	MIGRATION AND ORIENTATION	41
	3.1 Distances and altitude	41
	3.2 Orientation and navigation	43
	3.2.1 Use of visual cues	44
	3.2.2 Earth's magnetic field	46
	3.2.3 Infrasounds	48
	3.2.4 Odours	48
	3.2.5 Integration of cues	49
	3.3 Energy conservation during migratory flight	51
	3.3.1 Fuel	51
	3.3.2 Migration routes	52
	3.3.3 Use of tail winds	53
	3.3.4 Thermals and updraughts	55
	3.3.5 Formation flight	55
Chapter 4.	THERMOREGULATION	56
	4.1 'Normal' body temperature	56
	4.2 Torpor	57
	4.3 Thermoneutral range	60

CONTENTS

	4.4	Cold exposure	60
		4.4.1 Shivering thermogenesis	61
		4.4.2 Circulatory adjustments	64
		4.4.3 Respiratory adjustments	66
		4.4.4 Penguins in the Antarctic	67
		4.4.5 Nervous mechanisms	68
	4.5	Heat stress	69
		4.5.1 Dry heat loss	69
		4.5.2 Heat storage	70
		4.5.3 Brain temperature	70
		4.5.4 Evaporative heat loss	72
		4.5.5 Flapping flight	76
		4.5.6 Panting and respiratory gases	77
		4.5.7 Nervous mechanisms	78
	4.6	Birds compared with mammals	79

Chapter 5. OSMOREGULATION — 81

5.1	General considerations	81
5.2	Thirst and salt appetite	83
5.3	Drinking water and evaporative water loss	84
5.4	Physiological considerations with respect to environmental salinity	85
5.5	Physiological strategies in osmotic survival	85
	5.5.1 Sites of importance in salt and water balance	85
	5.5.2 Discovery of the physiological role of salt glands	89
	5.5.3 Salt glands	89
	5.5.4 The lower intestine as an integrator of renal and intestinal excretion	102
	5.5.5 The kidney	105
5.6	Behavioural aspects of osmoregulation	110
5.7	The gut	111

Chapter 6. THE REPRODUCTIVE SYSTEM AND ITS FUNCTIONS — 112

6.1	The gonads and reproductive tract	113
	6.1.1 The ovary	113
	6.1.2 The oviduct	115
	6.1.3 The testes and the male reproductive tract	116
6.2	The brain and pituitary gland	117
	6.2.1 Gonadotrophin-releasing hormone	118
	6.2.2 Follicle-stimulating hormone and luteinizing hormone	119
	6.2.3 Inhibitory steroid feedback	120
	6.2.4 Stimulatory steroid feedback	121
	6.2.5 Prolactin and its central control	122
	6.2.6 Gonadal steroids and sexual behaviour	123
6.3	Interactions with thyroid hormones	124
6.4	Interactions with the external environment	125
	6.4.1 Photoreceptors	125
	6.4.2 The biological clock	126
6.5	The ovulatory cycle	126
	6.5.1 The timing of ovulation	126

	6.5.2	Circadian rhythms and the 'open period'	128
	6.5.3	Plasma hormones and ovulation	129
	6.5.4	Some explanations of the 'open period'	130
6.6	Incubation and brooding		131
	6.6.1	Endocrine changes during incubation and brooding	132
	6.6.2	The initiation of incubation: progesterone or prolactin?	135
6.7	Moult		138

Chapter 7. THE ENVIRONMENT AND REPRODUCTION — 140

7.1	Breeding strategies		141
	7.1.1	Migration	141
	7.1.2	Delayed sexual maturation	142
	7.1.3	Non-annual breeding	142
	7.1.4	Tropical seasonal breeders	145
	7.1.5	Spring and summer breeding	146
	7.1.6	Autumn and winter breeders	150
7.2	Proximate factors initiating breeding		151
	7.2.1	Initial predictive information	151
	7.2.2	Supplementary information	153
	7.2.3	Synchronizing and integrating information	154
7.3	The mechanism of photoperiodic induction		155
	7.3.1	The critical daylength	157
	7.3.2	The role of circadian rhythms	159
7.4	Autonomous rhythms and scotorefractoriness		162
	7.4.1	Tropical opportunistic breeders	164
	7.4.2	Migratory species	165
	7.4.3	Autumn and winter breeders	165
	7.4.4	Scotorefractoriness: its role in autumn and winter sexuality	166
7.5	Absolute photorefractoriness		167
	7.5.1	Control by the central nervous system	167
	7.5.2	Dependence on daylength	168
	7.5.3	Role of the circadian system	169
	7.5.4	Some endocrine explanations	170

Chapter 8. APPLIED ASPECTS — 174

8.1	Domestication		174
	8.1.1	Chickens	174
	8.1.2	Turkeys	178
	8.1.3	Ducks and geese	179
8.2	Intensive farming		181
	8.2.1	Lighting patterns	181
	8.2.2	Broodiness	184
	8.2.3	Induced moult	185
	8.2.4	Stocking density and reproductive performance	185
8.3	Pollution		187
	8.3.1	Physical effects of oil contamination in birds	189
	8.3.2	Effect on the embryo	190
	8.3.3	Systemic effect of oil pollution	190
	8.3.4	Effect on juveniles	190

CONTENTS

	8.3.5	Responses in the endocrine system	191
	8.3.6	Impact of petroleum products on reproductive performance	193
	8.3.7	Incubation and breeding	194
	8.3.8	Embryonic and post-embryonic effect of pollutants	194
	8.3.9	Relevance of laboratory studies to studies in the wild	195
	8.3.10	Other pollutants	196
REFERENCES			199
INDEX			211

Acknowledgements

The authors are grateful to Professor W.N. Holmes, Dr A.J. Woakes and Dr N.H. West for reading parts of the manuscript and to Mrs V. Whitaker, Mrs C.M. Middleton and Mrs G. Green for typing the manuscript; to Mr R. Wheeler-Osman, Mr N. Day and Mr E. Armstrong for assistance with the graphical work and to Mr R. Wheeler-Osman, Mr R.K. Field and Ms D. Palmer for assistance with the photographic work. The authors are deeply indebted to Professor Ian and Mrs Nansi Chester-Jones of Sheffield who generously gave of their time to compile the index.

CHAPTER ONE

GENERAL INTRODUCTION

The impetus for this book arose from the recognition of the importance of birds as a focus of both amateur and professional scientific study. Unlike many other animals, birds are highly visible and their appearance and behaviour is, with few exceptions, universally attractive to man. They are thus a source of enjoyment—objects of beauty to be watched and wondered at by the amateur and professional scientist alike. As a class, birds are remarkably homogeneous in terms of their physical appearance and metabolism, reflecting the constraints imposed by the primary adaptation to flight. These adaptations include the evolution of effective homeostatic mechanisms and physical characteristics, such as lightness and feather insulation, which provide a degree of independence of the external environment equal to that enjoyed by mammals. All physiological and morphological adaptations reflect the evolution of patterns of behaviour and in this respect, birds are remarkably innovative. The evolution of diverse strategies of feeding, and of migratory, navigational and reproductive behaviour has enabled birds to occupy a very wide range of ecological niches and has contributed to the outstanding success of the Class Aves.

In addition to their aesthetic qualities, birds are endowed with easily studied morphological, behavioural and physiological adaptations which illustrate general principles of biological and practical importance. It is for this reason that birds are used as models in many areas of research, including cancer research, developmental biology, growth studies, reproductive behaviour, endocrinology, molecular biology and nutrition. It was the study of birds (notably the finches of the Galapagos Islands) which provided Darwin with some of the most important evidence for evolution. More recently, the work of Tinbergen and Lorenz on bird behaviour has made a major impact in the field of animal behaviour, which was recognized by the award jointly to them of the Nobel Prize for Medicine.

Some birds are of economic importance, either as a source of food (both eggs and meat) or as agricultural pests. In this context, a scientific study of birds assumes practical importance.

The contents of any book, and especially one designed as an introductory text with a limitation on size, rarely reflect in a totally comprehensive way the promise held out in the title. The reasons for this are usually numerous, not least amongst which is the need to balance informed opinion and the transmission of facts against the possibly more important task of drawing together a set of broad principles.

The ecological reasons for the diversity of bird behaviour and life-styles are covered in a separate companion volume to this (Perrins and Birkhead, 1983). *Physiological Strategies in Avian Biology* is not intended to be a comprehensive treatment of all aspects of the physiology of birds: rather it attempts, as the title suggests, to place the emphasis on physiological strategies which account for the success of birds. The text is designed to be self-contained and, as a result, to stand on its own, but its reading will be enhanced immeasurably by further reading of the references cited and also by the 'sideways' exploration of areas of avian biology not covered in this text. Further, it is presented in a way which is intended to satisfy a perceived dual purpose—that of providing a text for advanced undergraduates and postgraduates on the one hand while, on the other, being sufficiently readable to give an insight into the functional aspects of bird physiology to the informed amateur ornithologist. Whenever possible the authors have directed the reader to key review papers or books rather than primary source material but references to papers of this kind are made when this is considered necessary.

CHAPTER TWO

LOCOMOTION

Although flight is the major means of transport in birds, there are species which are variously adapted for life on water where an ability to swim and dive efficiently for food is of paramount importance. A few species have lost the power of flight altogether and have become extremely well adapted for aquatic or terrestrial forms of locomotion.

The study of locomotion in birds has an extra dimension compared with that in some other vertebrates, for many species of birds migrate twice a year over long distances, sometimes without stopping and sometimes at high altitude. An important factor associated with all forms of locomotion, but particularly with migration, is the distance that can be travelled for a given amount of fuel. This is obviously related to the mass of the bird, so the energy cost of transport, as it is called, is defined as the amount of energy required to transport one unit of body mass over one unit of distance (Schmidt-Nielsen, 1972).

Before getting involved in the details of locomotion, it is worth briefly considering three systems that are of central importance in the physiology of exercise in general, the respiratory, circulatory and muscular systems.

2.1 Respiratory system

The respiratory system of birds has a unique and complex structure (Figure 2.1). Air enters each lung by way of the main (primary) bronchus, which originates from the trachea. This primary bronchus runs the full length of the lung and gives rise to two groups of secondary bronchi, the medioventral group and the mediodorsal group, and to the posteriorly placed air sacs. Other, more anterior, air sacs originate from the medioventral secondary bronchi (Figure 2.1*b*). The two groups of secondary bronchi are joined together by tertiary bronchi (more commonly called parabronchi). From the lumen of each parabronchus there radiates, in the peri-parabronchial tissue, a meshwork of fine air capillaries which intertwine

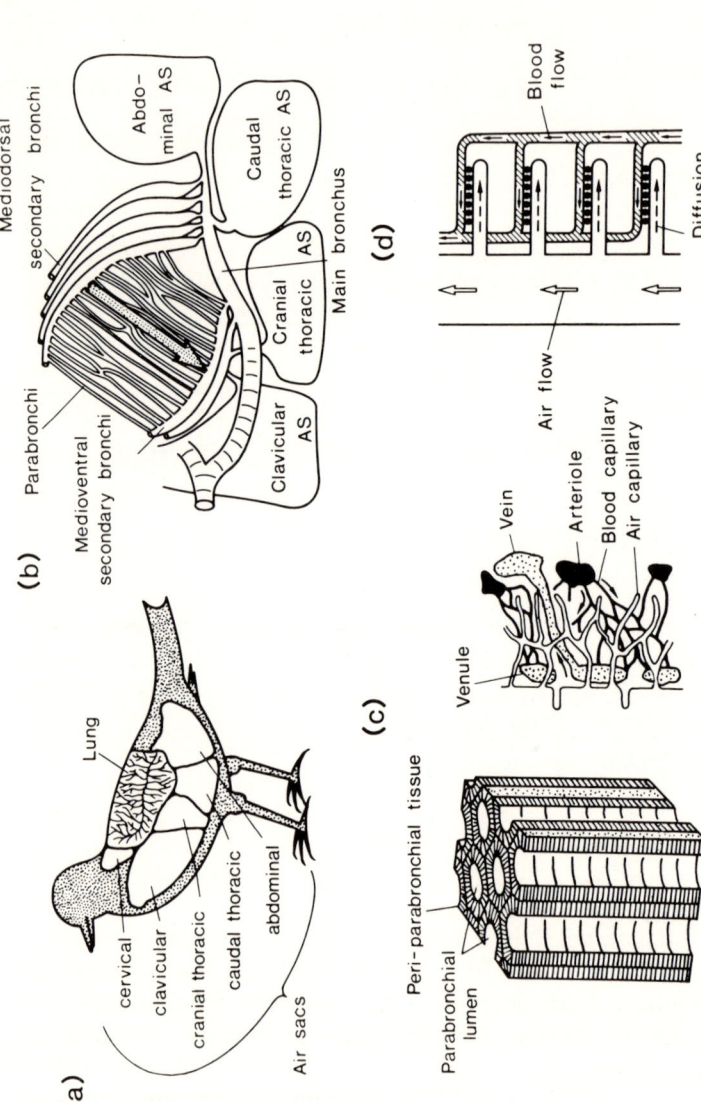

Figure 2.1 Diagrammatic representation of the respiratory system in birds. It consists of a pair of lungs and a number of air sacs which are connected to the lungs (*a*). Air passes through the lungs and into the air sacs during inspiration. Its passage through the lungs is in the same direction during both inspiration and expiration i.e. from dorsal to ventral secondary bronchi via the parabronchi—see arrow in (*b*). Surrounding the lumen of each parabronchus there is a dense network of air capillaries and blood capillaries, and this is where gas exchange occurs (*c*). Bulk air flow through the parabronchi and bulk blood flow through the arterioles and venules are at right angles to one another, giving rise to the cross-current system of gas exchange (*d*). (After Scheid, 1979.)

with equally fine blood capillaries. The blood capillaries arise from arterioles at the periphery of the peri-parabronchial tissue and themselves give rise to the collecting venules near the parabronchial lumen (Figure 2.1c).

Despite this complex arrangement of the lung, air flows in the same direction through the parabronchi (from mediodorsal to medioventral secondary bronchi) during both phases of ventilation (Scheid, 1979). Air then diffuses along the air capillaries where gas exchange occurs (Figure 2.1d). Bulk air flow through the parabronchi is at right angles to bulk blood flow through the arterioles and venules and this arrangement has been called a cross-current system. Elegant experiments by Scheid and Piiper (1972) demonstrated that this arrangement is more effective at exchanging gases (CO_2 in particular) than is the mammalian lung. The partial pressure of CO_2 (P_{CO_2}) in expired gas almost always exceeds the partial pressure of CO_2 in arterial blood in birds; this rarely occurs in mammals.

Although very little experimental work has been performed on birds it is probably safe to deduce, from the extensive work on mammals, that the respiratory rhythm is generated by groups of inspiratory and expiratory neurones in the brain stem. What we do know is that important sensory areas concerned with the control of the rate and depth of respiration are found associated with the major arteries and the lungs (see Scheid, 1982). Those in the lungs do not all respond to mechanical deformation, as they do in mammals. This makes sense, for the lungs of birds do not change volume to any great extent as the air passes through the lung and into the air sacs. There are, however, substantial changes in the partial pressure of CO_2 in the lung, and in ducks up to 80% of the sense organs in the lung respond to changes in the level of CO_2, being inhibited by relatively high CO_2. Thus as the bird inspires, the level of CO_2 in the lung decreases, activity in the sensory nerves in the vagus (Figure 2.2) increases and inspiration is terminated. These receptors seem to be concerned with the pattern of ventilation.

At the base of each carotid artery, close to the thyroid gland and innervated by the vagus nerve, is a small organ called the carotid body (Figure 2.2). This senses the partial pressure of gases in the arterial blood and when stimulated by low O_2 (hypoxia) and/or high CO_2 (hypercapnia) it causes an increase in ventilation. If these peripheral chemoreceptors are denervated, the ventilatory response to hypoxia is abolished and that to hypercapnia is reduced, but *not* completely abolished. There is evidence to suggest that centrally placed chemoreceptors, probably in the ventrolateral

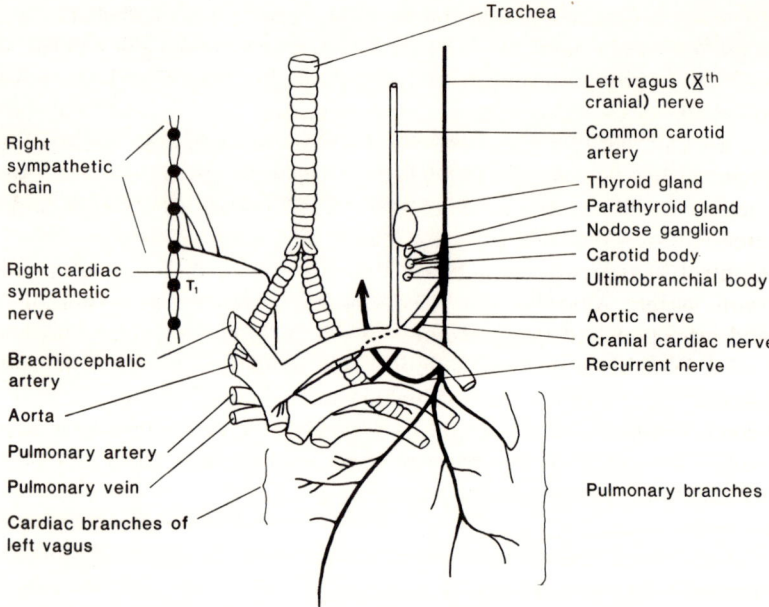

Figure 2.2 Diagrammatic representation of the left vagal (parasympathetic) innervation of the central cardiovascular system and lung and of the right sympathetic innervation of the heart in birds. T_1 indicates the first thoracic sympathetic ganglion.

region of the medulla oblongata, are sensitive to changes in the partial pressure of CO_2 (and/or pH) in the blood of ducks.

2.2 Circulatory system

The heart of birds is completely divided into right and left atria and ventricles and its contraction is initiated by a group of special cells localized in a region of the right atrium known as the sinuatrial node (see West *et al.*, 1981, for further details). Functioning of the heart is regulated by the autonomic nervous system. The parasympathetic system innervates the heart via branches of the vagus (Figure 2.2). Increase in activity of the vagus nerves to the heart causes the release of acetylcholine and a reduction in heart rate and possibly a reduction in the strength of contraction of the heart. Sympathetic innervation to the heart is via branches from the last few (up to three) cervical sympathetic ganglia (Figure 2.2)—those 'associated with the spinal nerve(s) contributing to the brachial plexus' (Cabot and Cohen, 1980). Increase in sympathetic activity causes the release of noradrenalin (maybe adrenalin) and an increase in both rate and strength of contraction of the heart. There is evidence, from the few birds that have been studied, that there is a continual (tonic) activity of both sympathetic and parasympathetic nerves to the heart of resting birds so that an increase in heart rate is often the result of an increase in sympathetic tone and a decrease in parasympathetic (vagal) tone.

Control of the peripheral blood vessels is by way of the autonomic nervous system. Amongst other things, this determines the distribution of blood in response to the level of activity of an organ, especially skeletal muscles. Whether a particular blood vessel constricts or dilates depends upon the type of receptor that is activated and the type of transmitter that is released. Receptors that respond to adrenergic nerve endings have been classified as α or β on the basis of their reaction to a series of related catecholamines (noradrenalin, adrenalin, isoprenalin). In the birds that have been studied α-adrenergic receptors are associated with constriction of peripheral blood vessels (vasoconstriction) and β-adrenergic receptors are associated with increased rate and strength of contraction of the heart and with vasodilatation of peripheral blood vessels. There may also be cholinergic nerve endings which have a dilator effect on peripheral blood vessels. The action of these neurotransmitters on specific receptors can be blocked by specific drugs. For example the β-adrenergic receptors are blocked by drugs such as propranolol whereas the receptors to acetylcholine are blocked by atropine.

Sense organs in the circulatory system of birds are found in the wall at the base of the aorta and are innervated by the aortic branch of the vagus nerve (Figure 2.2). These serve to prevent short-term changes in arterial blood pressure and are called baroreceptors. If there is an increase in blood pressure these baroreceptors cause a reflex decrease in heart rate and peripheral vasodilatation, and *vice versa* if blood pressure decreases. There is some controversy as to whether or not baroreceptor endings are present in the atria or ventricles of the avian heart (West *et al.*, 1981). Other sense organs which can have a profound effect on the cardiovascular system are the carotid body chemoreceptors, but only if ventilation is prevented from increasing. Under these circumstances stimulation of these receptors causes a reduction in heart rate and peripheral vasoconstriction. If ventilation does increase, as is usually the case, these cardiovascular changes are not apparent.

2.3 Skeletal muscles

Skeletal muscles are composed of fibres which may be classified, according to their contractile and histochemical properties, into three main groups. Slow, tonic fibres are found in specialized sites such as the anterior latissimus dorsi muscle which holds the wing against the body of birds. These muscle fibres do not exhibit a propagated action potential; instead each fibre is innervated by a number of end plates and shows a graded response to stimulation of different frequencies. They contract slowly and can remain contracted for a long time with little utilization of energy. Slow twitch (or type I) fibres do have a propagated action potential but they receive a low frequency pattern of stimulation and exhibit a slow rate of twitch contraction. According to the classical picture of the contractile mechanism in striated muscles, sliding motion (contraction) is generated through the interaction of actin and myosin filaments via the formation of cross-bridges. In these fibres the rate of cross-bridge turnover is slow. Thus, the frequency at which each cross-bridge must be reprimed with ATP is relatively low, so that tension may be maintained very economically (Goldspink, 1977). They use predominantly oxidative metabolism so they do not fatigue very rapidly and are well suited to perform sustained work, e.g. the postural muscles.

Fast twitch (type II) fibres receive a high-frequency pattern of stimulation, have a high turnover rate of actin–myosin cross-bridges and therefore show a fast rate of twitch contraction. Tension is maintained far less economically than in slow twitch fibres. Histochemically, two main sub-types of fast muscle fibres have been identified. Fast glycolytic (type IIb) fibres have few mitochondria and possess predominantly glycolytic (anaerobic) enzymes. They are, therefore, unable to supply ATP as quickly as they use it and they fatigue very quickly. Fast oxidative glycolytic (type IIa) fibres contain more mitochondria and have both oxidative and glycolytic enzymes. They are better able to supply the ATP required in the process of contraction and fatigue less rapidly than the type IIb fibres.

2.4 Energetics of flight

Man has been fascinated for centuries by the ability of birds to fly but has, until quite recently, been unable to obtain direct measurements of the energy expended during flight. It is not easy to obtain physiological data from a flying animal.

2.4.1 *Power output*

Not surprisingly, perhaps, a number of theoretical models have been presented which require minimum data to be inserted into simple formulae in order to predict the mechanical power required for flight (power output) of a bird at different flight velocities (V) (e.g. Pennycuick, 1969; Rayner, 1979). There are three major components to total power output (Figure 2.3): parasite power which is required to overcome the drag of the body and increases with increasing V, induced power which is required to support the weight of the body and decreases with increasing V, and profile power which is needed to overcome the drag of the flapping wings. Each downstroke of the wings accelerates air downwards and backwards, producing an upward and forward force on the bird. The air movements associated with each downstroke form a vortex ring (Figure 2.4) and the bird leaves a trail of such vortex rings in its wake. These rings have been visualized by training birds to fly through a cloud of neutrally buoyant helium bubbles. Induced power is that required to generate the vortex wake

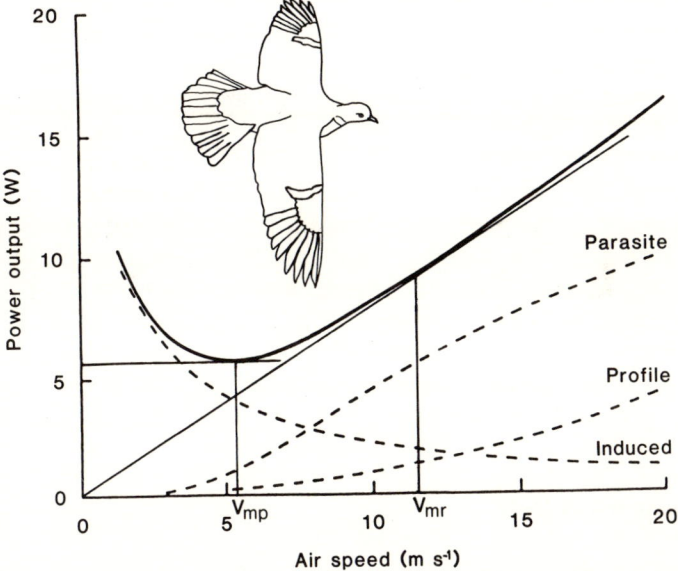

Figure 2.3 Predicted total power output/velocity curve for a pigeon of mass 0.333 kg during forward flapping flight. The three major components of this curve are indicated by the dotted lines. Also shown are the velocity at which power output is minimum (V_{mp}) and the velocity at which range is maximum (V_{mr}) for a given amount of fuel. (After Rayner, 1979.)

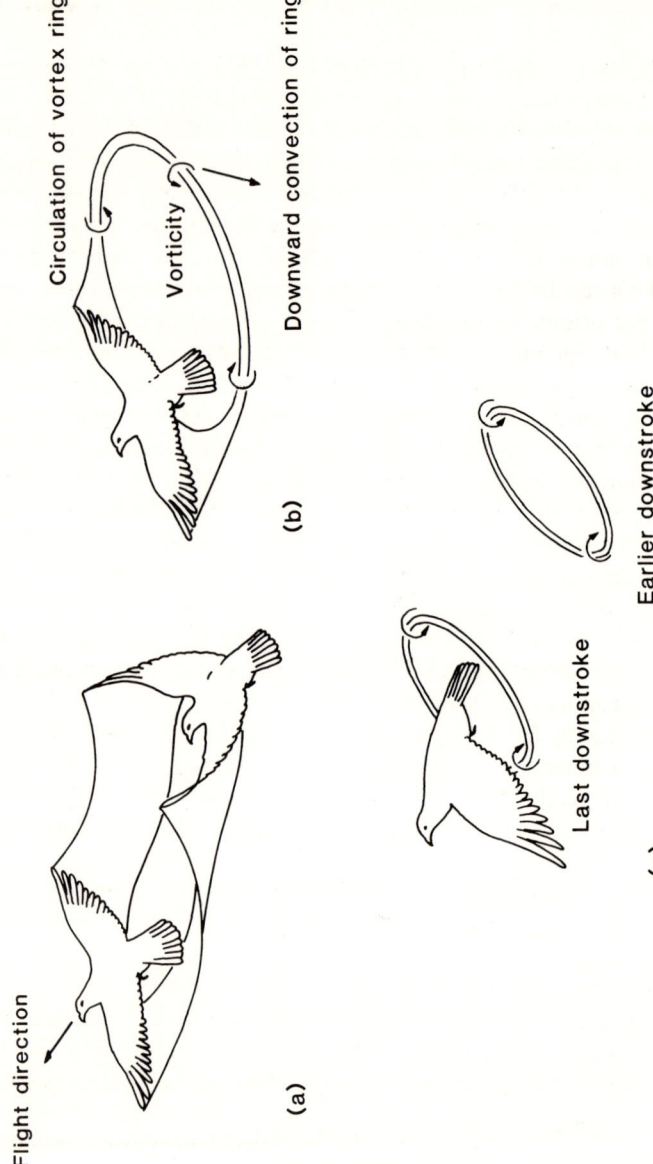

Figure 2.4 The vortex distribution generated by a single downstroke of the wings in a flying bird. (*a*) The vortex sheet shed during the first half of the downstroke. (*b*) The half vortex ring formed by roll-up of the vortex sheet shown in (*a*). (*c*) The complete, elliptic vortex ring, immediately after the start of the next upstroke. (After Rayner, 1980.)

and is calculated as the mean rate of increase of kinetic energy of the wake. Calculations indicate that induced power is high at low velocities and decays approximately in proportion to V^{-1}. Parasite and profile powers are easier to estimate and both increase in proportion to V^3. The resulting power output/velocity curve for a pigeon (Figure 2.3) is roughly U-shaped.

Thus, from this theoretical approach, the speed at which power output is lowest (V_{mp}) is just over 5 m s^{-1} for pigeons. This is not, however, the most economical speed, i.e. the speed that will give the maximum range (V_{mr}) for a given amount of fuel (the speed at which the energy cost of transport is lowest). This can be found by drawing a tangent from the power output curve to the origin. This velocity for maximum range (V_{mr}) is just over 11.5 m s^{-1} for pigeons, but because the power output curve is almost linear and coincident with the line from the origin over a wide range of velocities, any speed within 10–17 m s^{-1} will give a similar range. In other words, the energy cost of transport in a flying pigeon does not change significantly from 10–17 m s^{-1}. It would seem sensible for pigeons to fly at the maximum possible speed within this range when travelling long distances.

Values of maximum range velocity, obtained from theoretical power output/velocity curves, are close to observed mean air speed of a number of migrating birds, with the exception of the large (10 kg) swans (Alerstam, 1981). This exception is not too surprising as power required for flight is proportional to body mass $(M)^{7/6}$ which means that, as mass of the flight muscles is proportional to M, the power required per unit mass of muscle should increase by $M^{1/6}$. Calculations suggest that the power available per unit mass of muscle decreases by $M^{-1/3}$, indicating that there is a well defined upper weight limit for flying birds (Pennycuick, 1975). This appears to be approximately 12 kg. Thus, swans may not be able to produce sufficient power to fly continuously for extended periods at V_{mr}. From the observations of Whooper Swans swimming for part of their migration (see section 3.1), long-duration flapping flight, even at V_{mp}, may not be possible either, especially at the beginning of migratory journey when the bird is carrying a store of fat.

Theoretical considerations suggest that the ratio of lift:drag increases with increasing altitude, so that a bird should fly at a height where it can obtain oxygen fast enough to maintain V_{mr} (Pennycuick, 1975). As weight is lost during migration, the power required for maximum range velocity decreases, so that a bird should be able to climb higher ('cruise climb') as its flight progresses. However, as Alerstam (1981) points out, wind greatly affects birds' energy expenditure and it would be expected that they would fly at altitudes with the most favourable winds. With a tailwind, calculated

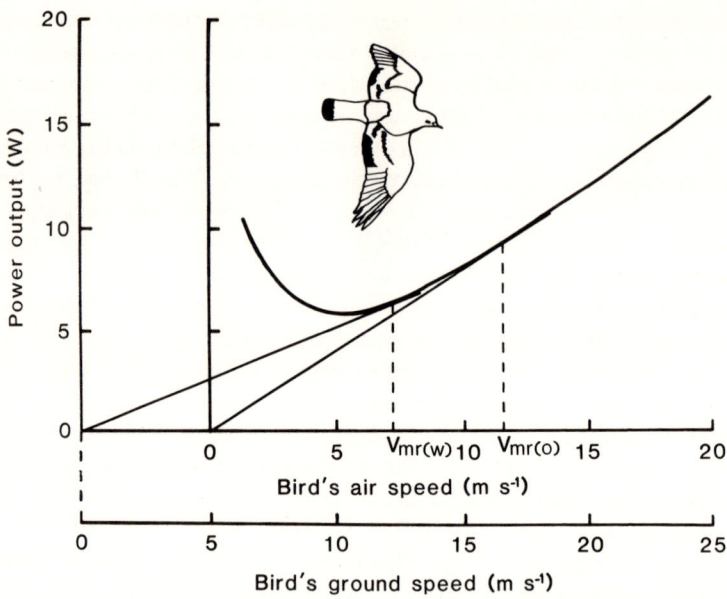

Figure 2.5 When flying with a tailwind of 5 m s^{-1}, the maximum range air speed, $V_{mr}(w)$, for a pigeon of mass 0.333 kg is less than maximum range air speed when the bird is flying in still air, $V_{mr}(0)$. The power output/velocity curve is the same as that in Figure 3.6.

V_{mr} is lower than that in still air (Figure 2.5), but because of the linear portion of the theoretical power output curves that have been constructed for birds so far (Rayner, 1979), the opposite effect of headwinds is not so easy to demonstrate graphically. Interestingly, though, observations indicate that birds do indeed have higher air speeds when flying in headwinds than in tailwinds.

2.4.2 *Power input and efficiency*

Unfortunately, useful and important though the theoretical considerations are, they do not tell us how much energy a bird actually expends when flying. They give an estimate of the mechanical (aerodynamic) power required for flight. As energy is lost in the biochemical processes of muscle contraction and in the mechanical linkages of bones, muscles and tendons, the total energy expenditure (or power input—P_i) is considerably greater than the estimated power output (P_o). The ratio P_o/P_i is known as the efficiency of the system and is usually reckoned to be 0.2–0.3 (but it could be higher), which means that up to 70–80% of power input is 'wasted' as heat. It is likely that efficiency varies with flight velocity and wing beat kinematics, but as yet we do not know how (Rayner, 1979). This means that values for flight velocity at which energy expenditure is lowest (V_{mp}) or at which energy cost of transport is lowest (V_{mr}) may not be accurate if they are based on a theoretical power *output*/velocity curve. The only accurate way of obtaining these important values is to construct a power *input*/velocity curve.

Some of the earliest values of metabolic rate during flight were estimated from the loss of weight during a long flight with the assumption that fat constitutes by far the major part of this weight loss. Using the calorific value of fat (39.75 kJ g^{-1}), it was calculated that a 19 g Blackpoll Warbler, flying at 6°–12°C, has an energy consumption of approximately 4.18 kJ h^{-1} (1.16 W or 3.5 ml O$_2$ STPD min^{-1}). Berger and Hart (1974), however, have criticized the method of measuring weight loss to determine total power used (power input) during flight as they believe that weight loss may exceed the production of metabolic water, i.e. respiratory water loss may be significant. In some studies the actual amount of fat that is used during a flight is estimated. Fat is estimated in a group of birds before a long flight, based on actual measurements from other birds of the same species. Fat content is then measured at the end of the flight. Such an estimate was performed on pigeons that flew a distance of 480 km. Carbon dioxide production was also measured in the same birds using doubly labelled water (D$_2$O^{18}), and both methods gave similar estimates of power input. Without a knowledge of the behaviour of the birds and of the prevailing meteorological conditions, it is impossible to apply these data to flight in general. They are relevant to the flight activity of the birds being studied only at the particular time of study. The most reliable way of obtaining power input is to measure it directly under controlled conditions.

The first direct measurement of power input of a bird during forward flapping flight was that of Tucker (1966), who measured oxygen uptake of Budgerigars trained to fly in a closed wind tunnel. In subsequent work Tucker attached a loose-fitting mask to the face of Budgerigars and Laughing Gulls, which enabled him to measure respiratory gas exchange and respiratory water loss under controlled flight conditions in open wind tunnels. A number of workers have since used basically similar techniques to measure a variety of physiological data from flying birds.

By measuring oxygen uptake directly and assuming that anaerobic metabolism makes an insignificant contribution to energy expenditure, it is possible to convert this to power input as 1 ml O$_2$ STPD s^{-1} = 20.1 W,

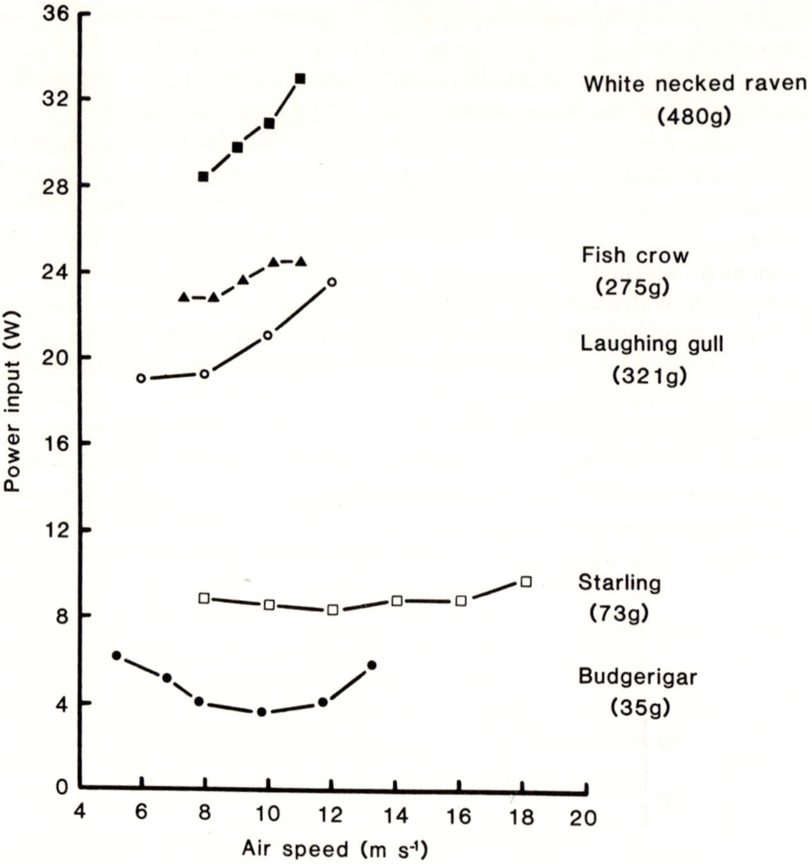

Figure 2.6 Power input at different air speeds for birds of different mass during horizontal flapping flight in a wind tunnel. (After Butler, 1981; Hudson and Bernstein, 1983.)

when respiratory quotient (RQ) is 0.8. Figure 2.6 shows power input (P_i) for a number of birds during level flapping flight at different velocities. Interestingly, only the budgerigar has the classical U-shaped curve as predicted by the theoretical models for power output (P_o) (see Figure 2.3). The three larger birds show a gradual increase in power input as velocity increases, although this is only slight for the Fish Crow. Perhaps they were not able to, or were not given the opportunity to, fly at lower velocities where power input would be expected to increase. Although of similar body mass, oxygen uptake during flight is different in the gull and Fish Crow, being higher in the latter. This is probably related to the facts that the wings are shorter and broader (lower aspect ratio) and wingbeat frequency is higher in the crow.

The Starling has a particularly flat power input/velocity curve, which may result from the change in body attitude that occurs during flight, and it means that minimal energy cost of transport for the Starling is at the highest sustainable flight velocity (cf. Figure 2.3). Assuming that the power output/velocity curve for the bird is U-shaped then, as indicated earlier, efficiency must change with different velocities. With this possibility it is surprising that the birds can fly no slower than $8\,\mathrm{m\,s^{-1}}$ and no faster than $18\,\mathrm{m\,s^{-1}}$. As with other birds that have been studied, the Starling has a more or less constant wingbeat frequency over a wide range of velocities. This is, no doubt, related to the fact that power developed by, and the mechano-chemical efficiency of, muscle are both maximal at particular contraction velocities (Goldspink, 1977). There is, therefore, an optimum wingbeat frequency, so it is changes in wingbeat amplitude that accompany changes in velocity (Figure 2.7). It is possible that upper and lower velocity limits are set by the starling not being able to beat its wings through a large enough amplitude to fly faster than $18\,\mathrm{m\,s^{-1}}$ or slower than $8\,\mathrm{m\,s^{-1}}$.

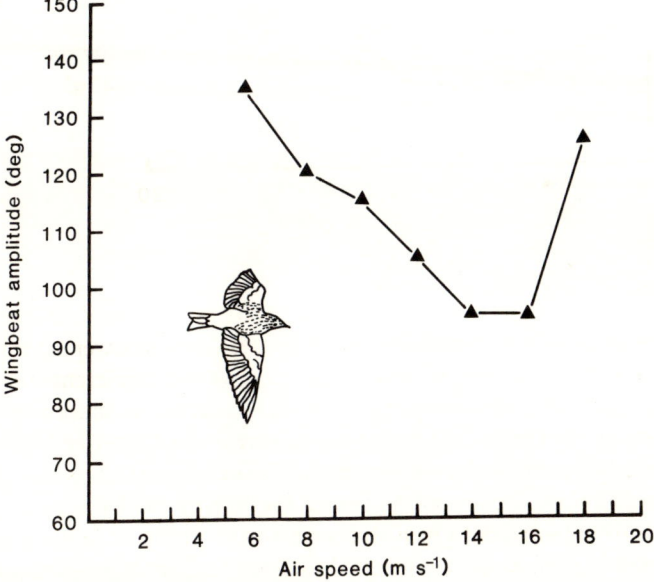

Figure 2.7 Wingbeat amplitude of the Starling, *Sturnus vulgaris*, during horizontal flapping flight in a wind tunnel, at different air speeds. (After Torre-Bueno and Larochelle, 1978.)

It is worth noting that although individual species of birds do not vary their wingbeat frequency to any great extent, there are large differences between species, with flapping frequency being inversely related to body size. A large bird, such as the Mute Swan, has a wingbeat frequency of about 2.7 Hz, whereas a small bird, such as the Redstart, beats its wings at about 15 Hz. The muscle fibres in large birds are very long with a number of sarcomeres in series. If the sarcomeres shortened at the same intrinsic rate as those of smaller birds, they would develop strain at a very high rate which could exceed the breaking force of the system. This may be one reason why they have lower wingbeat frequencies than smaller birds (Goldspink, 1977).

2.5 Factors affecting power input (oxygen uptake)

When an animal is exercising, measurements of oxygen uptake (power input) are normally affected to a limited extent by other factors as the exercise itself creates the greatest demand for the oxygen. There are, however, many factors which affect oxygen uptake in inactive animals. For diurnal species resting oxygen uptake is approximately 20% greater during the light phase of the daily activity cycle than during the dark phase and *vice versa* for nocturnal species. Oxygen uptake increases soon after a meal as a result of digestion and assimilation of food. This is known as specific dynamic action (SDA) and its magnitude will depend upon the type and amount of food that has been eaten.

Many birds expend large amounts of energy to maintain body temperature several degrees above ambient. The processes of thermoregulation are to be dealt with in Chapter 4 but it is worth mentioning here that there is a range of environmental temperatures, the thermoneutral range, over which resting oxygen uptake is unaffected. The thermoneutral range is different for different species but beyond this range, and particularly below it, there is an increase in oxygen consumption. Standard (or basal) oxygen uptake is defined as that measured in a resting, post-absorptive animal during a particular phase of its daily activity cycle at a thermoneutral temperature. This value may be up to 50% higher in winter than in summer and there is, on average, a 1% increase in standard oxygen uptake for each degree increase in latitude of the bird's natural habitat (Weathers, 1979).

One of the most important factors affecting oxygen uptake is the mass of an animal. It has long been known that smaller animals have a greater oxygen uptake per unit mass than larger animals (see Schmidt-Nielsen, 1983). If the logarithm of *total* standard oxygen uptake of a number of birds

LOCOMOTION

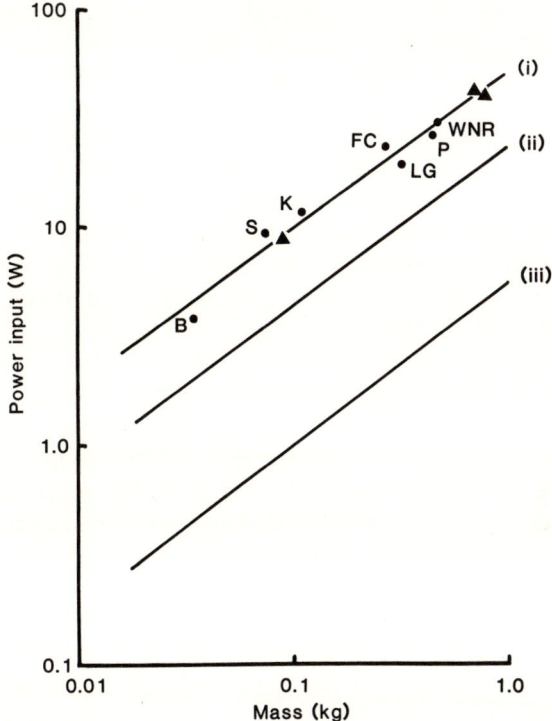

Figure 2.8 The relationship between power input, P_i, and body mass, M, for: (i) 7 species of birds during forward flapping flight in a wind tunnel, B = Budgerigar, S = Starling, K = Kestrel, LG = Laughing Gull, FC = Fish Crow, P = Pigeon, WNR = White-necked Raven. Minimum values of power input, P_{im}, were used to construct this curve, and $P_{im} = 50.4 M^{0.73}$ (data from Butler, 1981; Hudson and Bernstein, 1983); (ii) small mammals during maximum, sustainable exercise where $P_i = 22.6 M^{0.73}$ (data from Pasquis et al., 1970); (iii) resting non-passerine birds where $P_i = 5.5 M^{0.73}$ (data from Prinzinger and Hanssler, 1980). It is interesting to note that data from three species of flying bats (▲) lie on line (i) (data for bats from Butler, 1981.)

is plotted against the logarithm of their body mass, the points fall more or less on a straight line, as shown in Figure 2.8. The equation describing this relationship (known as an allometric relationship) has the general form

$$P_{is} = aM^b$$

where P_{is} = standard power input (oxygen uptake)
  a = mass coefficient (the 'intercept')
  M = mass of bird
  b = mass exponent (the slope of the line).

Although the exponent (b) is similar for all birds (approximately 0.75), it has, until recently, been thought that standard oxygen uptake is higher in passerine than in non-passerine birds i.e. that 'a' is greater in passerines (Lasiewski and Dawson, 1967). However, the equation presented by Prinzinger and Hanssler (1980) for standard metabolic rate for non-passerine birds weighing from 0.04 to 1.24 kg ($P_{is} = 5.5\ M^{0.73}$ W) is no different from those presented by other workers for passerines. These authors contend that if small non-passerines are included in the analysis, there is no difference in standard oxygen uptake between the two groupings.

The relationship between minimum power input (P_{im}) during level flapping flight of birds in wind tunnels and body mass has been found to be $P_{im} = 50.4\ M^{0.73}$ W (Figure 2.8). It is intriguing to note from Figure 2.8 that *minimum* power input during horizontal flapping flight is 2.2 times the *maximum* power input of exercising small mammals (i.e. up to 900 g). Despite the high rate at which oxygen is used during forward flapping flight, this form of locomotion is still relatively attractive as far as long migrations are concerned. The energy cost of transport is considerably less during flight than during walking or running (Schmidt-Nielsen, 1972) and the high rate of power input is manifest as higher velocities during flight.

2.6 Physiology of forward flapping flight

2.6.1 *Respiratory adjustments*

Not unreasonably, it has often been suggested that the structure of the avian respiratory system is related to the peculiarly high energy demands of flight. However, data from bats flying in a wind tunnel indicate that their oxygen uptake (power input) is similar to that of similar-sized birds during flight (Figure 2.8). Factors, other than the structure of the respiratory system, are clearly important in determining maximum oxygen uptake ($\dot{V}_{O_2max}$) during exercise in birds and mammals, at least at sea-level.

This topic has been dealt with in some detail by Weibel and Taylor (1981) working on mammals. They have proposed that the capacity of the gas transporting systems (respiratory and cardiovascular systems) is matched to the maximum demand that can be made by the muscular system. So, it could be argued that the factor which ultimately determines $\dot{V}_{O_2max}$ in healthy animals at sea-level is total mitochondrial volume (or activity of aerobic enzymes) in the exercising muscles. Certainly, increasing the partial pressure of inspired oxygen in rats has no effect on $\dot{V}_{O_2max}$. Direct measurements of barometric pressure on the top of Mt Everest (8848 m) indicate that it is higher than had been predicted, which allows a physiological explanation for how it was just possible for two climbers to reach the summit in 1978 without supplementary oxygen. The thought of birds flying at such altitudes is awe-inspiring, and it has been concluded that under such conditions of hypoxia the lung of birds with its high gas-exchange effectiveness does in fact contribute significantly to their high-altitude performance. Other factors, however, are also important (see section 2.6.3).

Unfortunately, the functioning of the respiratory system in flying birds has been studied in only a handful of papers. Direct measurements indicate that below an ambient temperature of 22°–23°C both Fish Crows and White-necked Ravens increase ventilation volume ($\dot{V}_I$) in proportion to oxygen uptake. This means that oxygen extraction (O_2Ext), which is the fraction of available oxygen removed from the inspired air, during flight is similar to that when the birds are at rest (Figure 2.9). In the crows, tidal

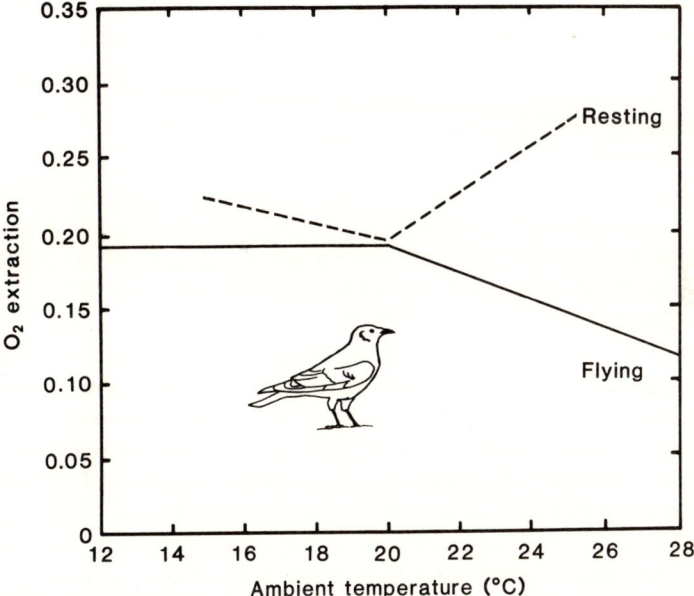

Figure 2.9 Oxygen extraction, the fraction of oxygen removed from the inspired air, at different environmental temperatures in Fish Crows, *Corvus ossifragus*, while at rest and during steady-state, horizontal flapping flight. (After Bernstein, 1976.)

volume during flight is almost double the resting value, whereas in the ravens the two values are similar. So in both instances the greater contributor to the increase in ventilation volume is respiratory frequency. Respiratory frequency and tidal volume in the Fish Crow and respiratory frequency in the Barnacle Goose are independent of flight velocity. This could mean that, like the Starling and the Fish Crow, oxygen uptake of the Barnacle Goose is largely independent of flight velocity. It is certainly interesting to note that the Budgerigar, with its distinctive U-shaped power input/velocity curve, also changes respiratory frequency with flight velocity in a similar U-shaped fashion.

Despite the apparent matching between ventilation volume and oxygen uptake at low ambient temperatures in the Fish Crow and White-necked Raven, there does appear to be an increase in effective lung ventilation, above that required by metabolic rate, during flight in Starlings at low ambient temperature: partial pressure of O_2 (P_{O_2}) in the air sacs increases and P_{CO_2} decreases. This effective hyperventilation results from the four times increase in tidal volume that occurs in this bird during flight. Simply multiplying tidal volume with respiratory frequency indicates that minute ventilation volume increases in proportion to oxygen uptake during flight. However, there is an increase in the fraction of tidal volume that passes through the lung, i.e. in (tidal volume – volume of large airways)/(tidal volume), during flight so that effective tidal volume increases by more than four times the resting value.

There is an increase in body temperature by approximately 2.0°C in all birds during flight at relatively low ambient temperatures, and this may explain the (effective) hyperventilation in Starlings. At ambient temperatures above 22°–23°C, body temperature of the White-necked Raven increases further during flight and in these birds and Fish Crows, ventilation volume increases progressively as ambient and body temperatures rise. As oxygen uptake does not change, there is a dramatic reduction in oxygen extraction (Figure 2.9). This increase in overall ventilation above metabolic demands is clearly related to temperature regulation via evaporative water loss and will be dealt with in more detail in the next chapter. It is important to note here that it will tend to cause a reduction in P_{CO_2} and an increase in pH in the blood (hypocapnia and alkalosis respectively), thus exaggerating the process already present, in Starlings at least, at low ambient temperature. It may also increase energy expenditure.

In the Fish Crow, though not in the White-necked Raven, the hyperventilation during flight at high ambient temperature results entirely from an increase in tidal volume. In other words, respiratory frequency stays constant. This is interesting because it means that there can be a constant relationship between wingbeat frequency and respiratory frequency. Ever since Marey's (1890) pioneering studies on bird flight, it has been known that lung ventilation may be co-ordinated with wing beating. In crows and pigeons there is a 1:1 correspondence between the two, whereas ratios as high as 5:1 (wingbeat frequency:respiratory frequency) have been reported for the Black Duck, quail and pheasant. These studies were on flights of only a few seconds' duration and it was concluded that co-ordination is not obligatory. However, during flights of up to 10 minutes' duration in a wind tunnel, pigeons show a very close relationship between the two activities. Wing beating occurs in bursts and although respiratory frequency is often either slower or faster than wingbeat frequency between bursts, there is always close co-ordination between the two when the wings are flapping (Figure 2.10). Almost immediately upon landing, the pigeons pant at a frequency close to the mean resonant frequency of the respiratory system,

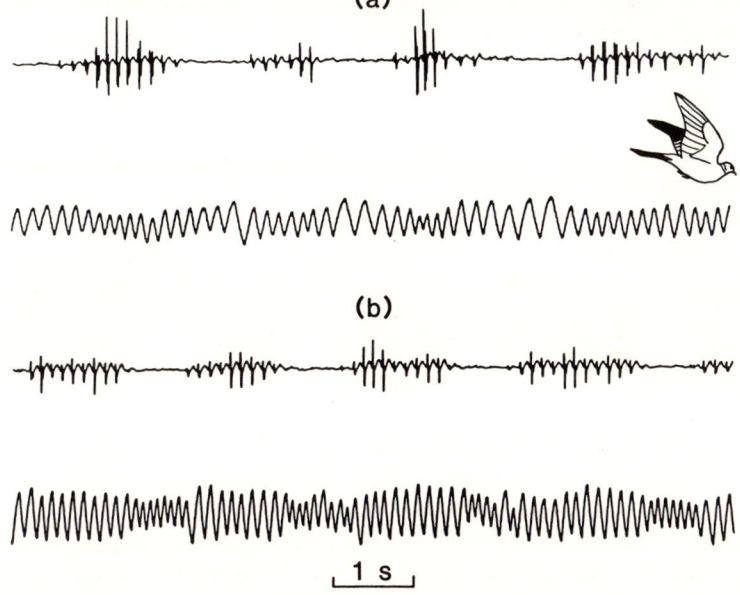

Figure 2.10 Traces from a pigeon, of mass 0.45 kg, during horizontal flapping flight showing changes in respiratory frequency associated with alternating periods of wing beating and periods of gliding, (a) 1 min after take-off when respiratory frequency decreased during gliding, and (b) 6 min after take-off when respiratory frequency increased during gliding. In each series the traces from top to bottom are: electromyogram from pectoralis muscle, respiratory movements (up on trace, inspiration). (Butler et al., 1977.)

i.e. 10 Hz. During flight, however, respiratory frequency is approximately 7 Hz, i.e. the same as wingbeat frequency.

With a 1:1 correspondence, it is possible to imagine how contractions of the flight muscles could assist respiratory air flow. With other odd-numbered ratios this is not so obvious. Nevertheless, a 3:1 correspondence has been found in the Barnacle Goose, and it is clear that the wingbeat is tightly locked to fixed phases of the respiratory cycle during flights of several minutes' duration. These phase relationships can be maintained even during transient changes in one of the activities. As Tucker (1968) comments: 'It is hard to believe that the contractions of the flight muscles have no influence on ventilation...'. This is obviously an area that merits further study. It is also worth mentioning that the pectoral muscles during flight may act as an effective blood pump as their contraction and relaxation affect the diameter of the blood vessels within both the muscles and the thoraco-abdominal cavity. Another interesting phenomenon noted in Barnacle Geese and in Canada Geese is that during flight the birds open their mouths during inspiration and close them during expiration. Perhaps aerodynamic factors reduce the energy cost of ventilation under these conditions.

2.6.2 *Cardiovascular adjustments*

The role of the various components of the cardiovascular system in presenting oxygen to the exercising muscles can best be described by Fick's formula:

$$\dot{V}_{O_2} = \text{H.R} \times (C_a O_2 - C_{\bar{v}} O_2)$$

where $\dot{V}_{O_2}$ = rate of oxygen consumption
H.R. = heart rate
S.V. = cardiac stroke volume
$C_a O_2$ = oxygen content of arterial blood
$C_{\bar{v}} O_2$ = oxygen content of mixed venous blood.

In pigeons flying at 10 m s^{-1} in a wind tunnel, steady-state oxygen uptake is 10 times the resting value. The respiratory system maintains $C_a O_2$ at slightly below the resting value but $C_{\bar{v}} O_2$ is halved, giving a 1.8 times increase in $(C_a O_2 - C_{\bar{v}} O_2)$. There is no significant change in cardiac stroke volume, so the major factor in transporting the extra oxygen to the muscles is the 6 times increase in heart rate. There is no significant change in arterial blood pressure during flight. Birds have larger hearts and lower resting heart rates than mammals of similar body mass. They also have a greater cardiac output (H.R. × S.V.) for a given oxygen consumption than similar-sized mammals. In other words $(C_aO_2 - C_{\bar{v}}O_2)$ is less in birds than in similar-sized mammals because $C_{\bar{v}}O_2$ is not reduced to such a low level. The higher cardiac output in birds may be an important factor in their attaining a higher $\dot{V}_{O_2 max}$ during flight than mammals when running. It is certainly interesting to note that bats have larger hearts and a higher blood-oxygen carrying capacity than other mammals of similar body mass (cf. Figure 2.8).

2.6.3 *High altitude*

There have been no physiological studies on birds flying at high altitudes (real or simulated) but work on inactive animals has indicated that adaptations of the cardiovascular system may be very important in the altitude performance of birds. Unlike Pekin ducks, Bar-headed Geese do not increase their haematocrit (packed cell volume of the blood) and

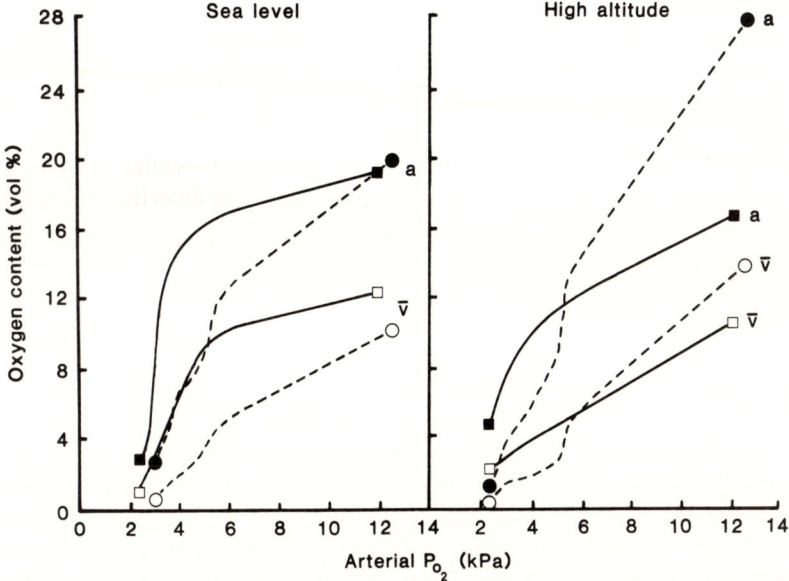

Figure 2.11 Relationship between oxygen content in arterial, a, and mixed venous, $\bar{v}$, blood and partial pressure of oxygen in arterial blood of Pekin ducks, *Anas platyrhynchos* (—-) and Bar-headed Geese, *Anser indicus* (———) at sea-level and after acclimation for four weeks to a simulated altitude of 5640 m. (After Black and Tenney, 1980.)

haemoglobin concentration when exposed to simulated high altitudes. This means that there is no increase in blood viscosity, thus preventing a possible increase in the energy cost of circulating the blood. It also means that there is no increase in the oxygen-carrying capacity of the blood. This is more than counterbalanced by the haemoglobin in the blood of the goose combining more readily with oxygen (i.e. having a higher affinity for oxygen). This enables the goose to maintain a higher C_aO_2 (and hence $C_aO_2 - C_vO_2$) at high altitudes than the duck (Figure 2.11). If not previously acclimated to high altitude, the ducks have a cardiac output of four times the sea-level value at a simulated altitude of 10 668 m; for the Bar-headed Geese the value is seven times that at sea-level. It is also apparent that the respiratory system of the Bar-headed Goose is able to maintain a very small difference between the partial pressures of oxygen in inspired air (P_IO_2) and arterial blood (P_aO_2) when at high altitude. At sea-level $(P_IO_2 - P_aO_2)$ is 7 kPa, whereas at a simulated altitude of 10 668 m it is a mere 0.5 kPa, giving proof of the effectiveness of the cross-current exchange system in birds.

Such low values of $(P_IO_2 - P_aO_2)$ are possible not only because of the anatomical arrangement of the lung, but also because of a large increase in ventilation, even though the demand for oxygen may not change. The hyperventilation, in response to lowered oxygen supply (hypoxia), results largely from stimulation of the carotid body chemoreceptors and it initially causes a fall in P_aCO_2 and an increase in pH_a; the bird becomes hypocapnic and alkalotic, as well as hypoxic. In a number of mammals (dog, monkey, rat, man) hypocapnia causes a reduction in cerebral blood flow; this is not so in ducks. Also, hypoxia causes a greater increase in cerebral blood flow in ducks than in dogs, rats and man. These two factors together, if present in other birds, will obviously be of great importance at high altitude.

2.6.4 *Future studies*

It is clear from this brief account of the physiology of flight that most of the useful physiological data have been obtained from a few studies using wind tunnels. Unfortunately, wind tunnels are not ideal because the bird has to fly in a restricted space, must tolerate the noise of the fan motor and is in an optically motionless environment. Also, it has been noted for the pigeon that the flight pattern is different compared with free-range flight (Butler *et al.*, 1977) and, unless very large tunnels are used, only the smaller birds can be studied. The early use of radiotelemetry did not demonstrate this technique to be a useful alternative. Because of the limited range of the transmitter, only very short flights (10–15 s) were possible. However, by imprinting Barnacle Geese on a human and training them to fly behind a truck containing the foster parent, it has been possible to record a few physiological variables, using an implantable telemetry system, from birds flying freely for several minutes (Butler and Woakes, 1980). With the rapid advance of electronic technology, it may soon be possible to record a number of important respiratory and cardiovascular variables from birds freely flying under these conditions.

2.7 Hovering, gliding, bounding and undulating flight

Forward level flight is one type of flying behaviour. Some birds, such as hummingbirds, hover at their food source; Pied Kingfishers hover for long periods over water into which they dive for food. When hovering, the bird remains stationary relative to the ground in otherwise still air. The air movements that support the bird's weight are generated entirely by the beating wings and induced power is large (see Figure 2.3). This is, therefore, the most energy-consuming form of flight, so the energy source (food) must be in plentiful supply. Oxygen consumption can be as high as $0.7 \, \text{ml} \, \text{g}^{-1} \, \text{min}^{-1}$ in small (3–8 g) hummingbirds (Berger and Hart, 1974). Most birds cannot (or do not) hover. Raptors and Petrels appear to do so but they are invariably flying slowly into wind or in an updraught. At the other end of the scale, gliding is a very economical form of flying. Direct measurements from Herring Gulls have shown that oxygen consumption is only twice the resting value when gliding compared with 6–8 times resting in the Laughing Gull during flapping flight. The increased oxygen uptake, above resting, when gliding is largely the result of the isometric contraction of the pectoralis muscle, which is necessary to keep the wing steady.

It is possible that birds which habitually glide have a special group of muscle fibres associated with this form of flight. In the majority of birds that fly, all the fibres in the pectoralis muscle are of the fast oxidative glycolytic type (IIa) and yet in the Herring Gull, 6% of the fibres are of the slow twitch type (I). These are probably used for gliding. This makes sense because, as already mentioned, slow muscles are more economical at maintaining isometric tension than fast muscles (Goldspink, 1977). The division of the pectoralis muscle of soaring birds into a large superficial part and a smaller, deep part may be further evidence of this functional separation of flapping muscle from gliding muscle. In albatrosses and Giant Petrels there is also a sheet of tendon forming part of the pectoralis muscle which, when the wing is fully extended, forms a locking mechanism, thus preventing it from rising above the horizontal position. This, no doubt, serves to reduce the energy cost of gliding even further in these birds.

Albatrosses, with their long, high-aspect-ratio wings, are the experts at gliding. These birds exploit the wind gradient over the surface of the sea. The lowest layers of air are retarded by contact with the sea but for the next few metres above the surface, wind speed increases quite rapidly with increasing height. By gliding into the wind an albatross extracts lift from the gradient and gains height (potential energy). When it reaches a height where the gradient becomes too weak, it turns and glides downwind, thus gaining kinetic energy during descent. This is known as dynamic soaring. Albatrosses may also make use of 'slope lift' resulting from the upward movement of air along a wave.

In between these two extremes (hovering and gliding) there are variations of forward flapping which affect the energy requirements. Using tilting wind tunnels it has been shown that, not unexpectedly, the energy requirements of ascending flapping flight are greater and those of descending flapping flight are less than horizontal flight. Gulls, crows and most birds of prey save energy by performing undulating flight. They gain height during a few wingbeats and then glide forward while losing height, and so on. Also, many small birds, with their short, low-aspect-ratio wings which are adapted for hovering and slow flight, can save energy during fast flight by 'bounding'. This consists of the birds folding their wings against the body for about half the time that they are flying and flapping that much harder for the rest of the time.

2.8 Take-off and landing

During take-off and landing, the pattern of wing beating is often different from that during sustained flapping flight. The pectoralis muscle of some birds is interesting in this respect. In pigeons, there are two types of fast fibre. There is a small, fast oxidative glycolytic type (IIa) that contains a high level of myoglobin, numerous mitochondria, high levels of fat, oxidative enzymes and lipase activity. There is also a larger fast glycolytic type (IIb) that has no fat, fewer and smaller mitochondria, low levels of oxidative enzymes but high levels of glycogen and glycogen-metabolizing enzymes. It has been suggested that the larger, type IIb, fibres (14% of total) are involved in brief bursts of activity such as take-off, rapid accelerations or sudden manoeuvres and use glycogen as a fuel, whereas the smaller fibres are involved in sustained flapping flight and metabolize mainly (but not necessarily exclusively) fat. Sparrows and Herring Gulls do not have any type IIb fibres in their pectoralis muscles, whereas chickens and pheasants have a high (85–90) percentage. The presence of type IIb fibres in pigeons may explain why the respiratory quotient (RQ) during relatively short flights is closer to 1 in these birds than in others that have been studied.

2.9 Terrestrial and aquatic locomotion

Although flight is the predominant form of locomotion in birds, it is by no means the only form. From studies on birds that have been trained to run on a treadmill or swim on a water channel, it seems that $\dot{V}_{O_2max}$ is very similar in those that are not primarily cursorial, e.g. Tufted Duck, Marabou Stork, Emperor Penguin, to that in running mammals of similar body mass (Butler, 1982). However, in the exclusively cursorial forms, such as the Emu, cockerels and rheas $\dot{V}_{O_2max}$ is considerably higher and close to the extrapolated line of $\dot{V}_{O_2min}v$. body mass for birds flying in a wind tunnel. As Prange and Schmidt-Nielsen (1970) point out, this means that a bird that can fly will have a higher $\dot{V}_{O_2max}$ during flapping flight than when running, but that in the Mallard Duck at least the ratio of $\dot{V}_{O_2max}$ during each form of exercise is similar to the ratio of mass of the muscles involved (pectoral and leg muscles). Thus, as suggested earlier, the total volume of mitochondria in the active muscles may limit $\dot{V}_{O_2max}$ at sea-level rather than any inadequacy of the cardiorespiratory system. The implication is that the mass of the leg muscles (or at least total mitochondrial volume in the muscles) of the cursorial birds is similar to that of the pectoral muscles of similarly sized birds that fly. This is somewhat theoretical, however, as the largest birds are exclusively cursorial. As well as $\dot{V}_{O_2max}$ being similar in running (bipedal) birds and running (quadrupedal) mammals, the energetic cost of transport is also similar, except for birds like penguins and geese which walk more awkwardly.

2.9.1 Walking/running

Penguins are, as we shall see later, extremely well adapted to aquatic locomotion, but some species spend a large amount of time walking on land. Two Antarctic species, the Emperor Penguin and the Adélie Penguin, walk long distances (100 km or more) every year to their rookeries. What is more, they may not feed while they are out of water. Nonetheless, the energy cost of transport is up to 75% higher for waddling penguins than it is for some other birds when running. Fortunately, the penguins are able to carry abundant fat which is more than sufficient. Less than 15% of the fat reserves of a large male Emperor Penguin, initially weighing 35 kg, are used in the process of walking 200 km to the rookery and back (Pinshow et al., 1976).

Although only the larger cursorial birds are likely to run for any long distances, a number of studies on the cardiorespiratory systems have been performed on running domesticated ducks and chickens, while there is only one study each on the turkey and the Emu. As might be expected the results are similar to those obtained from flying birds. An increase in heart rate is more important than a rise in cardiac stroke volume in increasing cardiac output and ventilation increases by a greater proportion than the rise in oxygen uptake (i.e. there is hyperventilation) as indicated by the animal becoming hypocapnic and alkalotic. These studies have given some information on the control of the respiratory and circulatory systems during exercise.

The increase in heart rate in turkeys results partly from the stimulation of β-adrenergic receptors either by the sympathetic nervous system itself or by way of an increase in circulating catecholamines. However, some cardioacceleration occurs at the highest running speeds ($> 3.5 \text{ km h}^{-1}$) in turkeys free of parasympathetic and β-adrenergic influences (Figure 2.12). The cause of this is unknown, although it could result from the increase in the rate of return of venous blood to the heart (venous return) caused by the contracting leg muscles. As might be expected, blood flow to the exercising muscles, via the ischiadic artery, increases during running, but so also does blood flow to the head, via the carotid artery, by 3.7 times and 2.3 times respectively. The latter is probably concerned with keeping the brain cool (see section 4.5.3).

Hyperventilation may be the result of the increase in body temperature that occurs during exercise. However, if ducks run in a cold environment and body temperature is prevented from rising, there are still signs of hyperventilation, although it must be pointed out that there was a metabolic acidosis (increased blood lactate and decreased pH) in these birds during exercise. This could well have caused the hyperventilation, for in chickens ventilation increases in direct proportion to gas exchange if body temperature is prevented from rising during moderate exercise. Although hyperthermia may be the cause of hyperventilation *during* running, it cannot be responsible for the rapid rise in ventilation that occurs with the *onset* of exercise in birds. Ventilation can still increase in running ducks

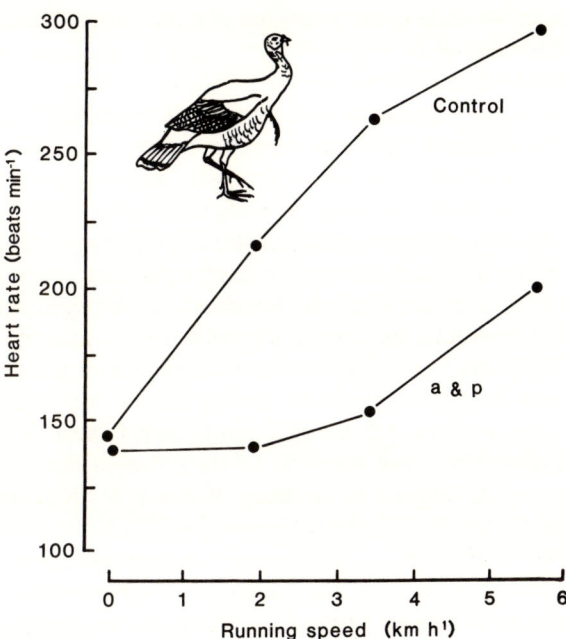

Figure 2.12 Changes in heart rate with running speed in untreated (control) turkeys, *Meleagris gallopavo*, and in those in which both parasympathetic and β-adrenergic influences on the heart have been blocked with drugs (atropine and propranolol respectively). (After Baudinette *et al.*, 1982.)

during exercise even in the absence of input from the CO_2 receptors in the lungs and if changes in P_aCO_2 and $[H^+]$ are kept to a minimum. This suggests that non-humoral factors are involved, such as muscle afferents and perhaps to a limited extent joint receptors. The increased rate of return of venous blood to the heart at the onset of exercise, caused by the contracting skeletal muscles, may itself stimulate receptors, perhaps in the heart, which causes an increase in ventilation. Such a mechanism has been proposed for mammals. It seems that arterial oxygen receptors play little part in the ventilatory response to running in domestic ducks breathing air, whereas they do provide a substantial proportion of the respiratory drive during running when the birds breathe oxygen-deficient (hypoxic) gas. This suggests that they are important in birds flying at high altitude.

Whatever initiates the adjustments in the cardiorespiratory systems at the onset of running, it appears that in birds, as in mammals, there is an increase in blood lactate at the beginning of exercise which may subsequently decrease or stabilize at the new level. This is presumably because the initial increased demand for energy is not matched by a sufficiently rapid increase in oxygen supply, so anaerobiosis occurs. At a certain level of exercise, lactic acid progressively accumulates in the blood. This is known as the anaerobic threshold, and exercise above this level cannot be maintained for long periods. The animal soon becomes exhausted.

Hormonal responses to exercise in birds have been poorly studied. The concentrations of glucagon and glucose in the plasma increase during running in domestic ducks, whereas insulin concentration is no different from that in control birds. Thus, unlike the situation in mammals, the rise in glucagon secretion during exercise cannot be due to low levels of insulin or glucose. From studies on Tufted Ducks, swimming at almost maximum sustainable velocity, plasma adrenalin and lactate are only twice the level at the lowest velocity. Although adrenalin is known to stimulate release of glucagon in ducks, it is not likely to be the cause of the release during aerobic exercise. Comparison of these two sets of data (the running domestic ducks and the swimming Tufted Ducks) may not be valid. The Tufted Ducks were trained over several weeks whereas it is not clear whether the domestic ducks were or not, and the level of fitness is important in these endocrine responses to exercise. There is an increase in the level of plasma corticosterone in unfit exercising ducks but this response decreases as the birds become fitter yet exercise at the same work load (Harvey and Phillips, 1982a). It would be interesting to know if the release of corticosterone occurs at a work rate equivalent to a given proportion of V_{O_2max}, as it does in man. If it does, then the results of Harvey and Phillips (1982a) could be explained in terms of V_{O_2max} being higher in the fitter birds. It appears that ACTH release is reduced in the fitter birds rather than the adrenal gland becoming less sensitive to ACTH.

2.9.2 Swimming

When walking or running, all birds that have been studied show a linear increase in oxygen uptake with increased velocity. Thus, the minimum cost of transport is at the highest sustainable speed (because when at rest the bird still consumes oxygen). However, for swimming Mallard and Tufted Ducks, oxygen uptake is relatively constant from speeds of 0.2–0.5 m s^{-1} but then increases very sharply as speed increases (Figure 2.13). This gives rise to a V-shaped cost of transport/swimming speed curve with the minimum cost at the transition speed (0.5 m s^{-1}). Interestingly, Prange and Schmidt–Nielsen (1970) found that mallards freely swimming on a pond

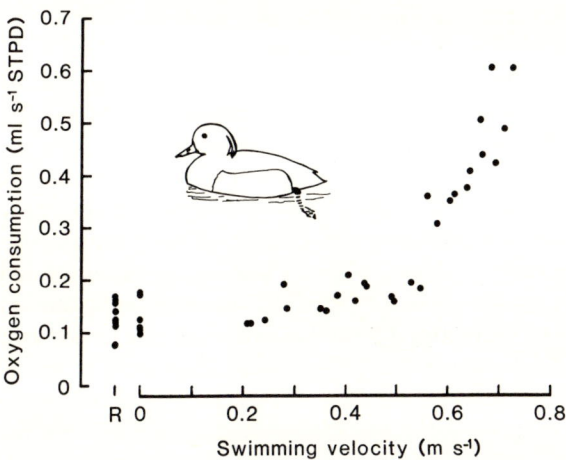

Figure 2.13 Oxygen consumption of Tufted Duck, *Aythya fuligula*, of mass 0.5 kg at different swimming speeds. R, rest with motor of water tunnel turned off. At $0 \, \text{m s}^{-1}$, motor of water tunnel was turned on. (Woakes and Butler, 1983.)

choose to swim at close to $0.5 \, \text{m s}^{-1}$. The change in heart rate with increased swimming speed is similar to that for oxygen uptake (Woakes and Butler, 1983), so there is a linear relationship between steady-state oxygen uptake and heart rate (Figure 2.14), just as there is in the Marabou Stork.

As a duck moves through the water surface it creates waves at its bow and stern and the wavelength of the waves is related to the speed at which it moves. When the bird reaches the speed where the wavelength of the bow wave equals the length of the body at the water line (hull speed), the body becomes trapped in the trough between the bow and stern waves. As hull speed is approached, resistance to further increase in speed (drag) increases almost exponentially. Hence the rapid rise in oxygen uptake of the swimming ducks at a certain speed. The mallards used by Prange and Schmidt–Nielsen (1970) had a waterline length of 0.33 m and a calculated hull speed of $0.71 \, \text{m s}^{-1}$. Maximum sustainable swimming speed of these birds was $0.7 \, \text{m s}^{-1}$.

Despite the nonlinear change in oxygen uptake with increasing swimming velocity, legbeat frequency and legbeat amplitude both increase linearly with a rise in swimming speed (Woakes and Butler, 1983). These data indicate that ducks' legs do not oscillate at a resonant frequency as suggested by Prange and Schmidt–Nielsen (1970). The gastrocnemius muscle of a range of birds consists predominantly of fast oxidative glycolytic fibres (type IIa) but also of about 20% of slow twitch fibres (type

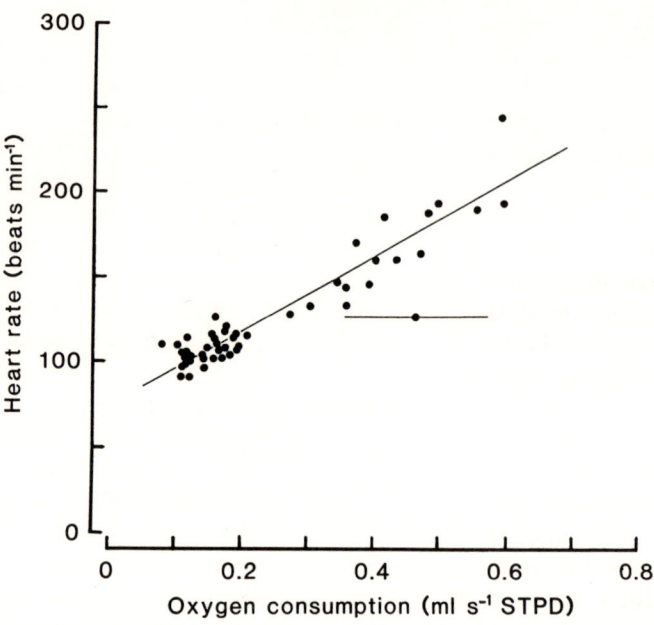

Figure 2.14 Relationship between heart rate and oxygen consumption of Tufted Duck (the same as in Figure 2.13) at different swimming speeds. The point with the horizontal bar is the mean oxygen consumption ($\pm$ S.E.) and mean heart rate at mean dive duration for the same bird. (Woakes and Butler, 1983.)

I). It is, therefore, somewhat different in composition from the pectoralis muscle in most birds. The supporting function of the legs is not present during swimming, as opposed to running, so it might be expected that, as in rats where the soleus muscle is not used during swimming, the muscles used during the two forms of exercise are slightly different.

Penguins swim under the water using their flippers as hydrofoils, whereas ducks use their feet as paddles. Penguins have a low drag coefficient, but they have to come to the surface to breathe. As they approach the surface the drag on the body increases and it has been suggested that leaping clear of the water altogether (porpoising) is energetically less costly than surfacing and encountering the highest resistance. It has been calculated that leaping is, in fact, energetically less efficient than swimming close to the surface up to a certain speed (cross-over speed), after which it becomes more efficient. The cross-over speed for the Adélie Penguin is approximately $2.5 \, \text{m s}^{-1}$, and these birds have certainly been seen to porpoise when swimming at high speeds.

2.9.3 Diving

Some birds are able to perform all forms of locomotion: fly, walk/run, swim/dive. An example is the Tufted Duck which obtains its food from the bottom of bodies of fresh water. It seems to prefer the freshwater mussel *Dreissena polymorpha*, but in its absence is omnivorous. Largely on the basis of work performed on domestic ducks that were involuntarily submerged, it was thought for almost 40 years that when birds (and mammals) dive there is, as a result of selective vasoconstriction, a reduction in blood flow to all tissues except the brain and heart, which continue to metabolize aerobically. The underperfused tissues metabolize anaerobically, producing lactic acid, thus saving oxygen for the oxygen-dependent regions of the body. Associated with the reduced blood flow to major regions of the body is a reduction in cardiac output resulting mainly from a large fall in heart rate (bradycardia). This bradycardia is the typical element of the response to involuntary submersion and is often taken as an indicator of the other adjustments taking place. The bradycardia and selective vasoconstriction result from the progressively more intense stimulation of the carotid bodies and central chemoreceptors as P_{O_2} in the blood decreases and P_{CO_2} increases. Thus, the oxygen conserving response takes 20–40 s to reach its maximum level and yet, with the exception of the larger penguins, birds do not usually remain under water, when diving voluntarily, for much longer than 60 s (Butler and Jones, 1982).

By recording heart rate (and sometimes respiratory frequency) by way of a small radiotransmitter, from naturally behaving Pochard, Tufted Ducks, and Humboldt Penguins, it was discovered that there is not an obvious, maintained reduction in heart rate during voluntary dives. Ducks and penguins dive repeatedly for extended periods without becoming exhausted. Preceding the first dive of a series there is an elevation in heart rate (Figure 2.15) and an increase in respiratory frequency in the ducks. Upon submersion, which appears to require much effort as the bird arches itself into the water, there is a transient bradycardia, but heart rate increases somewhat during the first few seconds of the dive and then remains at a more or less steady rate. The ducks have to continue paddling when under water and on occasions there is a 1:1 correspondence between heart rate and legbeat frequency. Surfacing occurs passively once leg-beating has ceased. By contrast, penguins glide under water and do not exhibit an obvious increase in heart rate before the first dive of a series (Figure 2.16), neither is it necessary for them to perform perceptible locomotor activity to remain submerged.

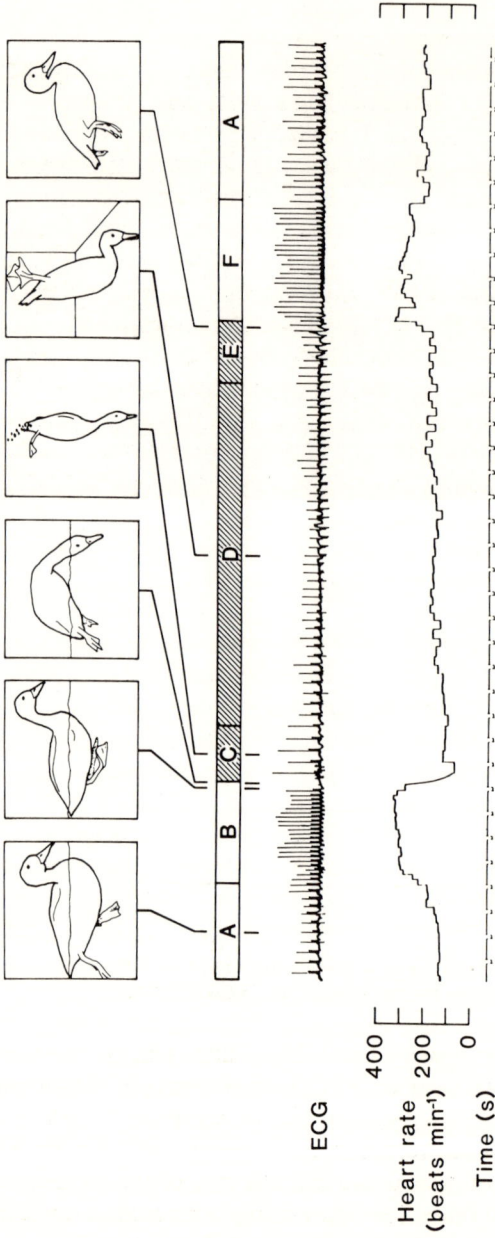

Figure 2.15 Relationship between the behaviour of a Tufted Duck and changes in heart rate during a spontaneous dive in a glass-sided tank, 1.55 m deep. From above, downwards: tracings of duck from cine film showing, from left to right, swimming, preparing to dive, moment of submersion, descending, feeding on bottom, surfacing (10 cm from surface); time periods of *A*, swimming on surface; *B*, cardiac acceleration before submersion; *C*, descent; *D*, feeding on bottom; *E*, surfacing; *F*, cardiac acceleration following surfacing; electrocardiogram (ECG); heart rate; time marker(s). The lines between the pictures of the ducks and the time boxes join coincident points in time (Butler, 1982.)

LOCOMOTION

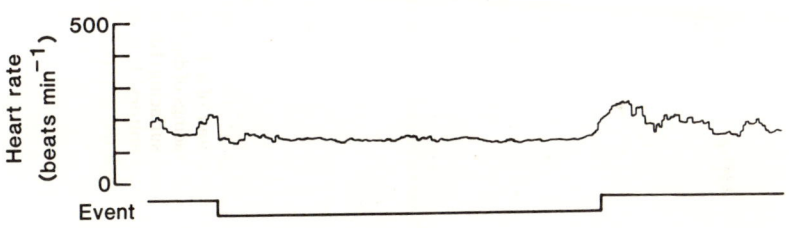

Figure 2.16 (*a*) Humboldt Penguin, *Spheniscus humboldti*, swimming on the surface of the pond with its eyes under water before submerging completely. (*b*) Heart rate of Humboldt Penguin (4.5 kg) before, during and after a voluntary dive. Duration of dive is indicated by the downward deflection of the event marker. (Butler and Woakes, 1984.)

Oxygen uptake has also been estimated for Tufted Ducks and Humboldt Penguins during voluntary diving. For the ducks, oxygen consumption at mean dive duration is 3.5 times resting and not significantly different from that measured at maximum sustainable swimming speed. On the basis of this estimated rate of oxygen usage and the size of the oxygen storage compartments in Tufted Ducks, aerobic metabolism could continue for 44 s with the duck actively swimming under water. The extra oxygen uptake during the anticipatory period before the first dive of a series increases aerobic dive duration by at least 7 s to a total of 51 s. This means that even ducks diving to a depth of 6 m in the wild, which takes an average of 28 s, do so aerobically, and have oxygen to spare at the end. Heart rate increases immediately upon surfacing largely, it has been suggested, as a result of renewed activity of inspiratory neurones in the brain stem and activation of the CO_2-sensitive receptors in the lung, both of which occur with the onset of ventilation. This ensures rapid replacement of the oxygen used during the preceding dive and enables the next dive to occur in quick succession. If mean heart rate and oxygen consumption during diving are plotted on the same graph as heart rate and oxygen consumption during surface swimming, then heart rate is lower during diving than during swimming at the same level of oxygen usage (see Figure 2.14). Diving heart rate is, on

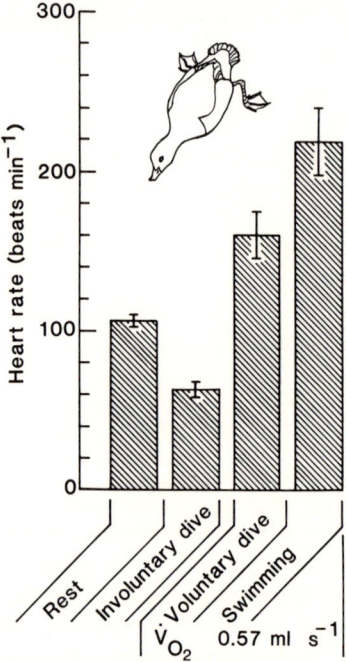

Figure 2.17 Mean heart rate ($\pm$ S.E. of mean) for 6 Tufted Ducks, *Aythya fuligula* (except during involuntary dives when 10 ducks were used), at rest on water, 15 s after involuntary submersion of the head, voluntary dives of 14.4 s duration and while swimming. Oxygen consumption ($\dot{V}_{O_2}$) at mean dive duration and while swimming was the same, 0.57 ml s^{-1} STPD. (Woakes and Butler, 1983.)

average, 1.5 times the resting value and 2.7 times the value after a similar period of involuntary head submersion (Figure 2.17).

On the basis of these results, it has been suggested that the circulatory adjustments during voluntary diving in ducks are similar to those during exercise in air inasmuch as the locomotory muscles as well as the heart and CNS receive an enhanced blood supply and sufficient oxygen for aerobic metabolism. The inactive muscles, viscera, kidneys and skin may well receive a reduced supply. The lower heart rate during diving could indicate, however, that selective vasoconstriction is more intense and that oxygen extraction by the active muscles is greater than when the birds exercise in air. During voluntary dives there would appear to be, in Tufted Ducks at least, a balance between the cardiovascular responses to involuntary submersion and to exercise in air with the bias towards the latter (Figure 2.17).

The balance can be tipped in the opposite direction if, for any reason, the duck is briefly unable to surface from a voluntary dive. As soon as the duck is aware that immediate access to air is not possible, there is a progressive bradycardia which, within 10–15 s, is similar to that seen during involuntary submersion (Butler, 1984). It would seem that the duck switches to the oxygen-conserving response. However, if ducks have to swim varying distances under cover in order to obtain their food, there is no initial lowering of heart rate when long distances have to be travelled, although heart rate does fall progressively during the longer dives after the first 5 s (Figure 2.18). These observations must be pertinent to ducks feeding under ice in the winter. It would seem that only if caught unawares does the animal immediately invoke its full oxygen-conserving response. When diving under ice, it is likely that ducks make a number of short exploratory dives initially in order to reduce their chances of becoming disoriented. If they have to swim long distances under water, they may shift progressively to the oxygen-conserving response.

For the Humboldt Penguins, neither heart rate nor oxygen uptake during diving are significantly different from the resting values. Calculations indicate that they can remain aerobic under water for 2.3 min. Observations on Chinstrap Penguins reveal that they dive to less than 45 m and for an average of 1.6 min. It certainly seems that penguins are able to dive and remain aerobic for longer periods than ducks. This is probably because they are more efficient at underwater locomotion, partly as a result of being almost neutrally buoyant. Biochemical studies suggest that some of the larger, more deeply-diving penguins, particularly the Emperor, King (both of which may dive to 250 m) and Adélie, may resort more regularly to

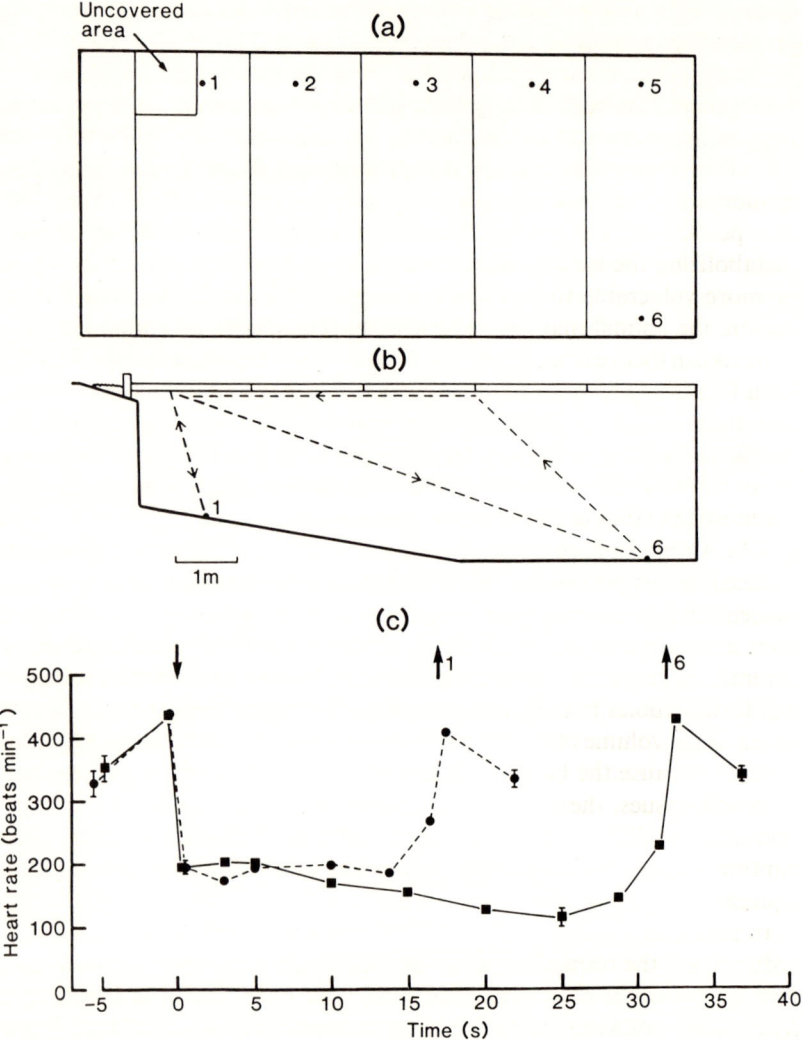

Figure 2.18 Plan (a) and side-view section (b) of an outside pond. The surface is almost totally covered except for one small area. Food is placed on the bottom of the pool at 6 different distances from the uncovered area (1–6 in (a)). (b) shows the route taken by the ducks to and from positions 1 and 6, while (c) shows the mean dive duration and mean (± S.E. of mean) heart rate for 6 Tufted Ducks while making these journeys. It should be noted that, for the longer, diagonal journeys, the ducks have to swim actively during the return as well as during the outward leg. (R. Stephenson and P.J. Butler, unpublished.)

anaerobiosis during feeding (diving) than the Royal, Rock-hopper and Gentoo Penguins. The Little Penguin, on the other hand, appears to be equipped predominantly for aerobic metabolism.

In general, the behaviour pattern of diving birds consists of using stored oxygen to metabolize aerobically during relatively short dives and then quickly replacing the oxygen at the surface before the next dive. This is more economical in terms of percentage of time spent feeding than if longer dives are performed, but proportionately longer is spent at the surface metabolizing the lactate. Also, animals in such an exhausted state would be more vulnerable to predators. However, if diving is extended for any reason, the animal may resort to the oxygen-conserving response.

Birds can fly to extraordinary high altitudes, but as already mentioned, it is the flightless penguins that descend to unusual depths in the sea. This may take them to regions where food is abundant, but it imposes physiological problems which are related to the fact that the animal submerges with some air in its body. For every 10 m descent, hydrostatic pressure increases by 1 atmosphere; thus at 100 m depth, total pressure is 11 atmospheres absolute (ATA). Air in a collapsible container, such as the lung, will have its volume reduced in proportion to the increase in surrounding pressure during descent and its pressure will be equal to ambient pressure. As nitrogen is inert it will accumulate in the blood and tissues and may cause narcosis. If ascent is too rapid to allow this nitrogen to leave the blood via the lung, it will form bubbles in the tissues (decompression sickness) and may lead to death. If the volume of the gas in the body is not reduced sufficiently during descent, because the body is not compliant enough, and if it is in contact with soft tissues, there will be a pressure difference across the tissue/air interface, as all body fluids will be at ambient pressure. This will cause rupture of blood vessels and pulmonary oedema (commonly known as 'the squeeze').

In penguins pneumatization of the bones and volume of the air sacs are reduced, but the respiratory volume is still large. It has been demonstrated that gas exchange occurs between blood and lungs of Adélie Penguins at hydrostatic pressures equivalent to 68 m depth. However, arterial P_{N_2} at this pressure is 3.5 ATA and only slightly above the value of 2.7 ATA recorded at 30 m equivalent depth. Thus, some impairment of gas exchange is present at the higher hydrostatic pressure and this may result from compression of the lung. Although P_{N_2} in the blood of the penguins after decompression was too low to cause bubble formation in the tissues or blood, it is known that symptoms characteristic of decompression sickness can occur in man after repetitive dives of short duration to shallow (20–

40 m) depths. More information is required to indicate how penguins, particularly Emperor and King Penguins, avoid the hazards of inert gas absorption and the squeeze during their feeding activity.

In order to feed effectively under water, penguins must be able to catch their prey (fish, squid and crustaceans) relatively easily. A recent study by Martin and Young (1984) has demonstrated that Humboldt Penguins have binocular vision and that the eyes are adapted to function in water. In air the eyes are myopic (they focus the image in front of the retina) because of the near-spherical lens (cf. fishes). When the penguin enters water the refractive properties of the cornea are almost negated by those of the sea water and the eyes become emmetropic (the image is focused on to the retina). Analysis of the visual pigments and oil droplets in the eye of the same species indicates that it should be capable of good wavelength discrimination at the blue-green end of the spectrum but that it has poor discrimination at the longer wavelengths, i.e. at the red end. This is consistent with the fact that this species of penguin is associated primarily with the deep coastal and oceanic waters of its S. American coastal breeding sites. These waters are classified as 'blue water' types in which there is relatively little light at the red end of the spectrum. The lack of red sensitivity in these penguins parallels that of fish which inhabit bodies of 'blue water', so the visual system of Humboldt Penguins, at least, is completely adapted to function in the aquatic environment.

CHAPTER THREE

MIGRATION AND ORIENTATION

Migration is one of the wonders of the animal kingdom. On the basis of natural selection it must confer reproductive advantages upon those that practise it and hence it must have contributed to the success of the Class Aves. As Aidley (1981) points out, migration may be an alternative to dormancy for some birds, e.g. those that spend the summer in the Arctic and move south for the winter. At more southerly latitudes, migration may take some species into areas of less competition. For example, five species of swallows and martins (Hirundinidae) migrate northwards, from Africa, for breeding in the summer whereas 31 species remain within the African continent. Aidley speculates that the total number of migrating hirundines is 'appreciably greater' than that of the resident African species.

There is a great deal of variation in the migratory patterns of different birds, both between and within species. The Arctic Tern migrates from its Arctic or subarctic breeding grounds to winter in the Antarctic, while tropical species of birds may 'migrate' merely a few kilometres as the seasons change. Within-species variation is illustrated by the Fox Sparrow. A population from Alaska spends the winter in southern California while a population from northern British Columbia winters in Oregon. In addition, there are sedentary populations in southern B.C., northern Washington and Vancouver Island. Migratory distances may vary from year to year. The lesser redpoll, which breeds in N. England, tends to winter in S. England if the birch seed crop is good but moves further south to continental Europe if the crop is poor.

3.1 Distances and altitude

Long distances are not always covered in a single flight, but are usually broken into a series of shorter flights. Most songbirds travel at night, beginning 0.5 to 1 h after sunset and covering 300–600 km a night. However, one individual does not usually fly each night and may require 3–

4 weeks to complete a 3 000 km trip. There are also migrants which perform similar, relatively short flights, but only during daytime. Longer flights are common among shorebirds and waterfowl, during which the birds apparently travel continuously, day and night. Some of the largest flying birds may not, in fact, fly continuously for aerodynamic reasons (see section 2.4.1). One flock of Whooper Swans, being tracked migrating N.E. over the Baltic Sea, landed on the water and swam for 2 h before taking off again (Alerstam, 1981).

In terms of the non-stop distances travelled by some migrating birds, the human marathon is like a stroll to the corner shop. Even the staggering of a man to the top of Mt. Everest without supplementary oxygen is not as impressive as the flying of migrating birds over the top of the Himalayas. Two classical examples of long distance, non-stop migrations are those across the North Sea, a distance of up to approximately 750 km which is covered in 12–15 h in good weather, and those across the Gulf of Mexico, which is a distance of approximately 1000 km and takes 12 h, on average. We now know that much more impressive feats are achieved and examples of some routes that are routinely used by migrating birds are shown in Figure 3.1. Greenland Wheatears often travel directly to France or the

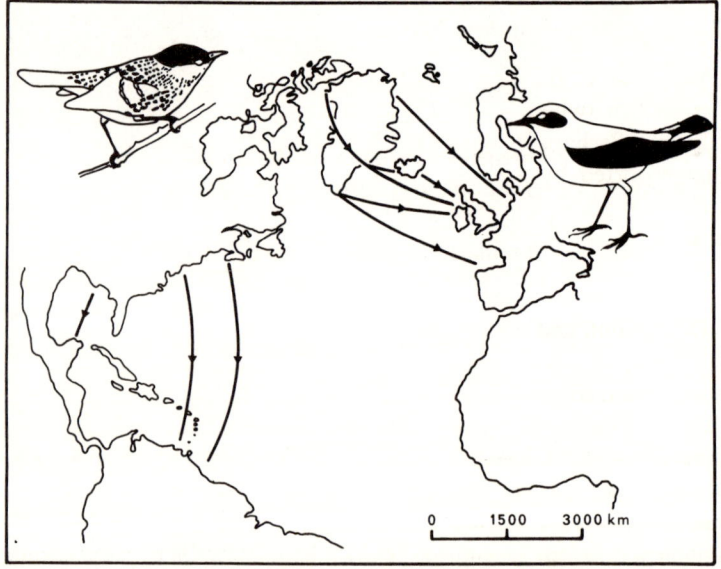

Figure 3.1 Examples of non-stop migration routes in the North Atlantic that are regularly used by birds. (After Alerstam, 1981.)

Iberian peninsula, which is a distance of approximately 3 500 km, on their way to Africa. The Blackpoll Warbler is one of many species that travels directly from N. to S. America across the Atlantic Ocean. This is a distance of 3000–4000 km. Outside the N. Atlantic other non-stop flights are just as impressive. Black Brent Geese fly about 4800 km across the Pacific Ocean from Alaska to Mexico, and some New Zealand cuckoos fly 3500 km to the Solomon Islands or Samoa. Not all non-stop flights are over water. Lesser Snow Geese fly from James Bay in N. America to the coast of the Gulf of Mexico, a distance of 2700 km, in less than 60 h.

It is not easy to obtain accurate information on the altitude at which birds fly during migration. Alerstam (1981) has presented some examples from reliable studies. Most passerines migrating at night travel below 2000 m with maximum heights lying between 3000–6300 m. Generally, migrating birds fly below clouds, as illustrated by a study on swifts. Under clear skies, average maximum altitude is 2300 m whereas during cloudy conditions it is 700 m. Of course the higher the altitude, the lower the temperature and the lower the partial pressure of oxygen (P_{O_2}), although the concentration of oxygen is the same as at sea-level. The physiological implications of these facts will be discussed later. It is worth mentioning here that on 9th December 1967 a flock of about 30 swans (probably Whooper Swans) was located by radar, near the island of Tiree off the W. coast of Scotland, at an altitude between 8000–8500 m where the temperature was $-48°C$. Bar-headed Geese have been observed flying at altitudes up to 9000 m (where P_{O_2} is 6.9 kPa) during their migration across the Himalayas. Perhaps the most reliable and the most impressive observation is that reported by Alerstam (1981), when a passenger and an aircraft captain saw small flocks of curlew-like birds. They were due north of Frankfurt, West Germany, at an altitude of 10 000 m above sea-level.

3.2 Orientation and navigation

The successful completion of a migratory trip depends to a large extent on the ability of the bird to travel in the correct direction. As we shall see later, a number of meteorological factors may influence direction over relatively short distances, but the bird must always be aware of the direction of its ultimate destination. Or must it? True navigation has been demonstrated convincingly in but a few species (Emlen, 1975). When experimentally displaced 750 km from their normal migration route (i.e. transported from Holland to Switzerland), young Starlings on their first migratory flight flew in an unchanged compass direction and therefore ended up far from the population's normal wintering ground. Some adults behaved in a similar fashion, at least initially, whereas others returned successfully to the normal wintering area, thus displaying true navigation. Studies of this type, whether displacement is experimental or natural, rely on reasonable recapture success. Displacement has been performed with the birds in cages and orientation behaviour has been studied. Unfortunately the results are somewhat confusing (Emlen, 1975). Perhaps such large-scale displacements rarely occur naturally. Even if they do, the fact that many of the young displaced Starlings mentioned above adopted the new wintering areas, and returned to them in later years, indicates that the lack of true navigation in juveniles may not always be disastrous.

From these, and similar studies, Able (1980) has concluded that most migrant birds employ some form of direction and distance orientation (vector navigation) on their first migration. Even though they may be able to correct for displacements by the wind, there is no strong evidence that they are capable of true, or bi-coordinate, navigation to a specific wintering ground. Previous experience appears to be necessary for the return to specific sites in subsequent seasons. The possible mechanisms by which this is achieved have been the subjects of many studies, and some of these mechanisms will now be considered.

3.2.1 *Use of visual cues*

Strange as it may seem, landmarks appear to play little role in migrating and homing behaviour (Figure 3.2*b*), and even celestial cues, the sun and the stars, may provide more simple information than was once thought (Keeton, 1981). Theoretical considerations as well as much experimental data indicate that the sun acts as a compass, merely allowing birds to maintain a bearing that is itself determined by some other means (Emlen, 1975). The 'mere' ability to maintain a bearing using the sun is possible only if the bird can compensate for the sun's apparent movement across the sky as the earth rotates. This was demonstrated to be so by the classical studies of Gustav Kramer (see Kramer, 1957). In other words, birds possess an internal clock and if this clock is artificially phase-shifted, for example by 6 h, the bird (starling or pigeon) initially orients at 90° to its normal flight direction (3.2*a, b*). From these 'clock-shift' experiments, there is no evidence that these birds use the sun for bi-coordinate, or true, navigation (Emlen, 1975). The sun navigation hypothesis (see Matthews, 1968) has thus fallen

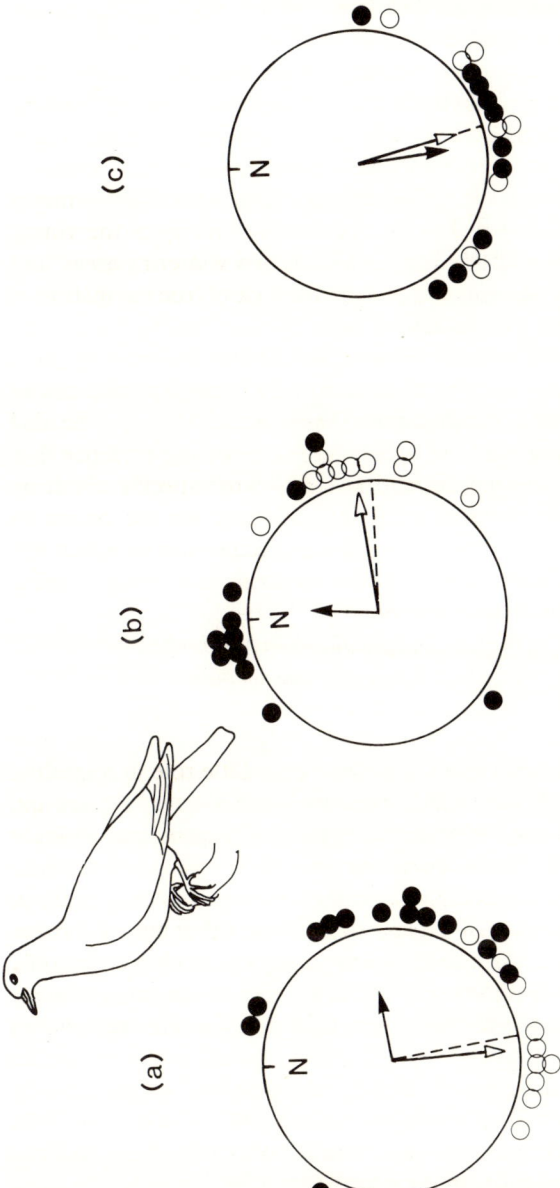

Figure 3.2 Direction in which pigeons fly to disappearance after they have been exposed to a 6 h advanced light/dark regime. (a) Experienced pigeons released on a sunny day at a distant site. The mean bearing of the phase-shifted birds (●) is approximately 90° to the left of the controls (○). (b) Experienced pigeons released on a sunny day at a site less than a mile from home where the landscape should be completely familiar. Again, the mean bearing of the phase-shifted birds is approximately 90° to the left of the controls. (c) Experienced pigeons released on a totally overcast day at a distant, unfamiliar site. Both phase-shifted and control birds head for home which suggests that in the absence of the sun, pigeons use cues that do not require time compensation. In all diagrams, north is indicated by N, and the home direction by a dashed line reaching the perimeter of the circle. The bearing of each individual bird is shown by the symbols outside the circle. The mean directions are shown by the arrows (⬆ phase-shifted, ⇧ controls) and the longer the line the tighter the clumping of the bearings. (After Keeton, 1981.)

into disfavour. It is perhaps worth mentioning, however, that pigeons *are* capable of discriminating between locations on the basis of a difference in sun-arc altitude and that an ability to do this is a component of the sun navigation hypothesis. It appears that the stars also provide compass information alone, but data from the Mallard and from the Indigo Bunting indicate that for some birds at least, time compensation is not important. These animals use the *patterns* of the stars to determine directions.

Despite possessing these remarkable abilities, it is not necessary for birds to use them in order to orient. Homing pigeons can fly homeward from distant, unfamiliar release sites under totally overcast skies (Figure 3.2c), and studies involving radar tracking demonstrate that overcast skies have no discernible effects on orientation or flight performance of nocturnal migrants. When flying within or between clouds, birds do not maintain a given direction as precisely as those flying below the clouds, but they are by no means disoriented (Able, 1982). Under these conditions it appears that non-visual cues are being used.

Before discussing non-visual cues, it is worth mentioning the possible role of polarized light in orientation. Cardiac conditioning experiments have indicated that pigeons can detect polarized light, provided the light falls on the central or lower part of the retina, the so-called yellow oil droplet field. The same technique has been used to demonstrate that pigeons can also detect UV light in the wavelength 325–360 nm. These observations raise the interesting possibility that birds can use the polarized UV light in the sky the same way that bees can (von Frisch, 1967). If so, then they may be able to use a derivative of the sun compass on partially overcast days when the sun itself is hidden from view but some blue sky remains. The geometric relationships between the position of the observer, the plane of polarization of sunlight and the position of the sun allow the sun's position to be computed. It has been demonstrated that pigeons can orient themselves to linearly polarized light and that the migratory orientation of the White-throated Sparrow, a nocturnal migrant, is predictably influenced by changes in the axis of sky light polarization at dusk. It is possible, therefore, that birds are able to make use of the navigational information contained in the changing patterns of sky light polarization with time.

3.2.2 *Earth's magnetic field*

The view that the earth's magnetic field could provide non-visual cues has had a chequered history for over a century but has recently gained new favour. The spontaneous nocturnal activity (*Zugunruhe*) of caged robins is oriented in the appropriate seasonal direction even when there are no visual environmental cues. If the birds are placed in a large steel chamber, which greatly reduces the total intensity of the local magnetic field, the ability to orient correctly disappears. What is more, the orientation of the robins can be predictably changed by altering the direction of the magnetic field using Helmholtz coils around the test cage (Merkel and Wiltschko, 1965). Attaching bar magnets to their backs causes free-flying homing pigeons to become disoriented under overcast skies, while under similar conditions, pigeons with Helmholtz coils attached to their heads fly off in different directions, depending on the direction of the artificial magnetic field. There is also evidence which indicates that unusual magnetic fields can slightly disrupt pigeons' orientation under sunny skies. In other words, there may be some interaction between the magnetic and solar cues rather than a switch from one to the other.

A possible sensory basis for these observations became apparent with the exciting discovery that there is, in the heads of pigeons, the ferromagnetic mineral magnetite (Fe_3O_4). Perhaps even more exciting is the fact that exposing pigeons to a procedure that should randomize the alignment of any magnetic particles in their head (called degaussing) affects their orientation and flight under overcast skies. Thus the magnetite found in pigeons 'may be the physical basis of magnetic field sensitivity' (Walcott, 1982). So far, three major sites of iron-containing tissue have been located in the heads of pigeons: the Harderian gland in the orbit, the base of the beak and the bony ledge just ventral to the olfactory nerves and olfactory lobes of the brain. There does not appear to be any iron-containing material in the pecten of Ring-billed Gulls and pigeons. Electrophysiological and biochemical data indicate that, in the pigeon, the pineal gland could be part of the magnetic compass, and as the pineal gland is also the location of a biological clock in birds (see section 6.4.2), it has been suggested that it is the site of integration of the magnetic compass/solar-clock system.

During level flight the head of a pigeon is held very stable and its posture is such that the visual projection of the boundary of the red and yellow oil droplet fields of the retina coincides with the natural horizon. In other words, the pigeon appears to view the area below the horizon with its red oil droplet field, while the sky is viewed with the yellow field. It has been suggested that the boundary of the oil droplet fields serves to stabilize the bird's head with respect to the visual horizon and that this is important for the functioning of the solar and magnetic compasses.

3.2.3 Infrasounds

In 1969, Griffin suggested that birds might use low-frequency sounds (infrasounds) of <1 Hz for navigation. Infrasounds are not audible to humans but they can travel thousands of kilometres without attenuation and so could provide navigational information. It has since been demonstrated that homing pigeons are very sensitive to infrasound and that they can detect the lowest frequency that could be produced in the investigators' laboratory, 0.06 Hz (Kreithen, 1978). The sensory system appears to be the same as that used to detect sounds in the human audible range and it has been proposed that homing pigeons could use the Doppler shifts induced by their flight speed to localize infrasounds. Thus, birds may locate their position on an acoustic map derived from sources of infrasound, such as those produced by winds blowing over mountains. Also, infrasound could indicate the position of hazardous or beneficial weather conditions.

3.2.4 Odours

To complete the list of senses that may be used by navigating birds, Papi *et al.* (1972) have proposed that olfaction plays some role in the homing of pigeons, as sectioning the olfactory nerve or plugging the nostrils impairs their homing performance. (The latter probably also has some effect on lung ventilation). It is suggested that pigeons become familiar with the odours close to their home loft and associate windborne odours with the direction of the wind carrying them to the loft, so obtaining an olfactory map of the area. Thus, if the direction of the wind entering the loft is altered, by deflector plates, the homing direction of the pigeons when released at distant sites is affected in a manner predicted by the olfactory hypothesis (Figure 3.3). Odours, and magnetic stimuli, encountered during the outward journey from the loft also appear to influence the orientation direction at distant sites. In further support of the olfactory navigation hypothesis is the fact that the homing performance of starlings deprived of olfactory information is also very much reduced. However, workers in other countries have experienced difficulties in repeating some of the experiments performed by the Italian group (see Able, 1980; Keeton, 1981; Papi, 1982 for discussions) and studies using wind deflectors could be interpreted in terms of the reflection of light cues by the deflector panels. Nonetheless, there appears to be agreement that odours do provide information that

MIGRATION AND ORIENTATION

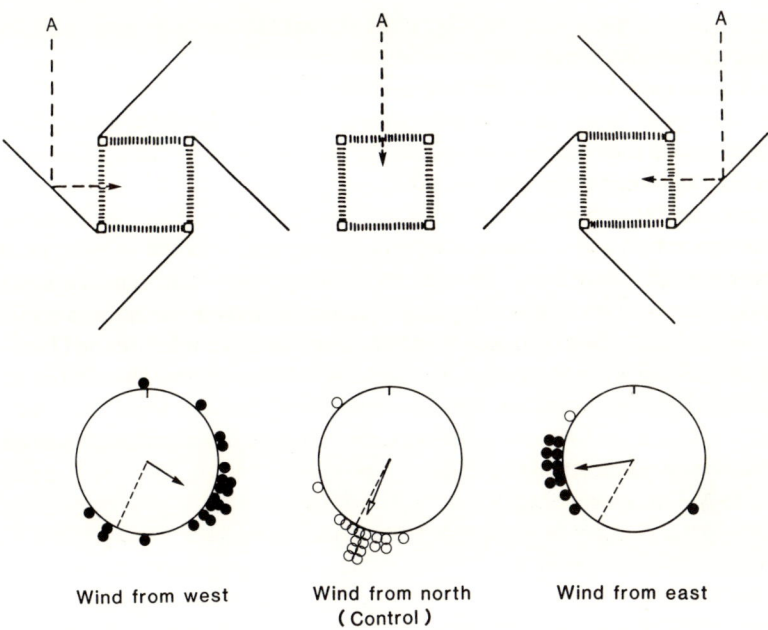

Figure 3.3 Effect of altering the direction of the wind entering the loft on disappearance direction of pigeons released north of home. Top: all three lofts (control loft in centre) have walls that allow free flow of air. Winds from the north, presumably carrying odour A, enter the control loft from the north, but they enter the other lofts from the east and west because of the deflector plates. Bottom: bearing of pigeons released from north of home. The control birds orient properly (i.e. southward) towards home whereas the other birds orient more easterly or more westerly depending on the direction of the deflected wind. Symbols as in Figure 3.2. (After Keeton, 1981.)

may be used by birds, particularly pigeons, during navigation. It is claimed, for these birds, that atmospheric odours are, in fact, a necessary component of the navigational 'map' system (Wallraff, 1983).

3.2.5 *Integration of cues*

With the discovery of more orientational cues and the probability of their being used during migration in birds, the question arises as to how all the various information is processed by the bird. Keeton (1981) presents an account of some of the tests that have been performed to investigate this fascinating area. It is clear that early experience is important in determining how, if at all, birds use particular cues. If unable to see the clear night sky during the weeks preceding their first autumnal migration, Indigo Buntings never learn to use the star compass. First-flight pigeons require both sun and magnetic cues (and, maybe, odours) for accurate orientation, yet if pigeons are raised without ever having seen the sun, they are able to orient accurately when released under totally overcast skies. Young pigeons raised permanently under a 6 h slow photoperiod orient normally. They learn that the 'morning' sun is in the south etc. When moved to a normal photoperiod, however, they orient at 90° to the control pigeons. It appears from these data that the sun compass must be calibrated against some other more fundamental 'reference' cue. By shifting the geomagnetic field, by way of Helmholtz coils, at the nests of Pied Flycatchers during incubation and nestling periods, Alerstam and Hogstedt (1983) produced evidence which they interpret as indicating that the earth's magnetic field serves as the primary compass cue.

The physiological condition of the birds seems to be an important component in the way they respond to the various cues. By manipulating photoperiod it is possible to bring some Indigo Buntings into autumnal condition at the same time as others are in spring condition. When tested under a spring sky in a planetarium the 'autumnal' birds orient southward whereas the 'spring' ones orient in the opposite direction. The possible involvement of hormones in this is indicated by the observation that orientation of the White-throated Sparrow in cages can be affected by altering the pattern of administration of prolactin and corticosterone.

It is clear from the studies on navigation and orientation that the senses in birds are far better than was once imagined. It was not long ago that we thought birds could not detect sound below a frequency of 200 Hz and that they could not see ultraviolet light. Not only have these myths been exploded but it now seems that birds can detect the earth's magnetic field and that they have an acute sense of smell. It appears that not all of the cues available to avian orientation systems are used at any one time; some are temporarily redundant. The manner in which these cues are integrated in different species of birds under different environmental condition is the problem which will continue to tax the ingenuity of the students of avian navigation. Navigation is, of course, only one component of the process of migration. The physical fitness of the birds is equally important if they are successfully to complete their migratory journey. For those that travel long distances without stopping, flight must be efficient, as the only available fuel is that which is 'on board'.

3.3 Energy conservation during migratory flight

This section could equally have been headed 'Flight economics', for it is concerned with the ways in which birds make the most economic use of their flying apparatus, particularly during migration. One simple way is to carry an energy-rich fuel.

3.3.1 *Fuel*

Fat, which is stored in preference to carbohydrate, produces 40 kJ of energy for each gram that is fully oxidized. For carbohydrate, the value is 17.5 kJ g^{-1}. The fuel ratio (F) is the fraction of take-off mass of migratory bird that consists of fuel (fat) and this seems to be related to the distance of non-stop flying. Smaller birds about to perform long, non-stop migrations may have a fuel ratio of 50%, whereas for short or middle distance migrants F is 13–25%. In a number of these latter birds, e.g. Chaffinches, there is often a larger fuel ratio in spring than in autumn. This may be related to inclement weather in spring and poor feeding conditions en route. On the other hand, several species of shore bird are fatter in autumn than in spring, which may be associated with a longer migration route in autumn. Larger birds do not have such large fuel ratios as smaller birds, because with too much extra weight they cannot attain the muscle power for maximum range (P_{mr}) (Pennycuick, 1969).

Even in smaller birds the increase in body mass, resulting from the deposition of fat prior to migration, is accompanied by an increase in the size (hypertrophy) of the pectoralis muscle so that sufficient power can be generated to cope with the larger body mass. This increase in muscle mass appears to be the sole factor involved in the rise in total aerobic capacity prior to migration. The mass specific aerobic capacity of the pectoral muscle of the Gray Catbird, as indicated by citrate synthase activity, is among the highest reported for skeletal muscle, but it does not change during pre-migratory fattening or in relation to the accompanying muscle hypertrophy (see Dawson *et al.*, 1983). The activity of β-hydroxyacyl-CoA dehydrogenase almost doubles during pre-migratory fattening. This suggests that fatty acid oxidation is enhanced, thereby sparing carbohydrate stores. The significance of this is that studies on endurance training in humans have demonstrated a relationship between fatigue and the depletion of muscle glycogen stores. Also, animals that have undergone endurance training rely less on glycogen than untrained subjects, at moderate work rates.

As a bird uses up fuel during a flight its mass decreases. This means that the power required to overcome induced and parasite drag forces also decreases and that the bird would have to reduce its speed progressively during the flight in order to keep near its maximum range velocity (V_{mr}). The changes required would be large, but if achieved they would allow a bird with an initial fuel ratio of 50% to fly 2.5 times the distance on 1 g of fat at the end of the flight (assuming the fat is almost all consumed) compared with soon after take-off (Pennycuick, 1969).

3.3.2 Migration routes

As indicated above, the migration route of a bird may not be the same in spring and autumn. The shortest distance between two points on the earth's surface is along the great circle, but to follow this requires a continual

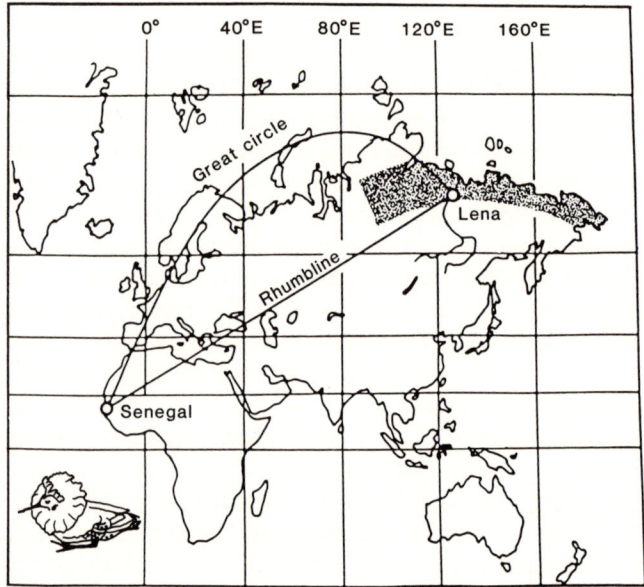

Figure 3.4 Great circle and rhumbline routes for Ruffs, *Philomachus pugnax*, migrating between the river Lena in E. Siberia and Senegal. The stippled area indicates the breeding range in E. Siberia. The routes are plotted on a Mercator map projection and areas close to the poles are disproportionately large. (Alerstam, 1981.)

change of compass heading. The path of constant compass heading, the rhumbline, intersects all longitudes at the same angle. Ringing studies indicate that Ruffs, which originate in E. Siberia, follow close to the great circle on their way to Senegal in the autumn (Figure 3.4). This distance is about 10 000 km, whereas the rhumbline route is 18% longer. In spring, however, they probably travel along a route between the great circle and the rhumbline because of a lack of resting sites, as the ground is frozen, along the great circle route (Alerstam, 1981).

3.3.3 Use of tail winds

Migration routes often relate closely to following winds, particularly in sea-birds. The Great Shearwater breeds in the Tristan da Cunha group of islands in the S. Atlantic. It then migrates in a clockwise direction into the northern hemisphere, initially to Newfoundland, and then across to western Europe. In doing so it takes advantage of the trade winds in the southern hemisphere and the westerlies in the north. The waters off Newfoundland, which are rich in food, are also reached by some birds, e.g. Fulmars, flying from Europe across the N. Atlantic against the westerly winds (Alerstam, 1981). Thus, sea-bird migration is not always directed downwind. In contrast, a fascinating example of the profound influence of wind direction on the destination of migration is that of the Redwing Thrush. Depending on the direction of the wind, it migrates from Scandinavia in the autumn to either Britain and W. Europe or to the E. Mediterranean and Black Seas. In other words, a redwing may spend the winter in Britain one year and in the USSR the next. Generally speaking, wind speed is less at low altitude than it is at high altitude. There is some evidence, particularly from the Chaffinch, to suggest that migrating birds may drift in strong winds at high altitude and compensate for any displacement from the migratory route at lower altitude where the winds are weaker.

From the foregoing account it is not surprising that the onset of migration is affected by weather conditions. Dense migration occurs with following winds and fair weather, although other factors such as falling temperature may be important, particularly in late autumn migrants, irrespective of wind direction. Weather conditions may change during migration and if they become too adverse, birds tend to land at the earliest opportunity (see Alerstam, 1981 for fuller account).

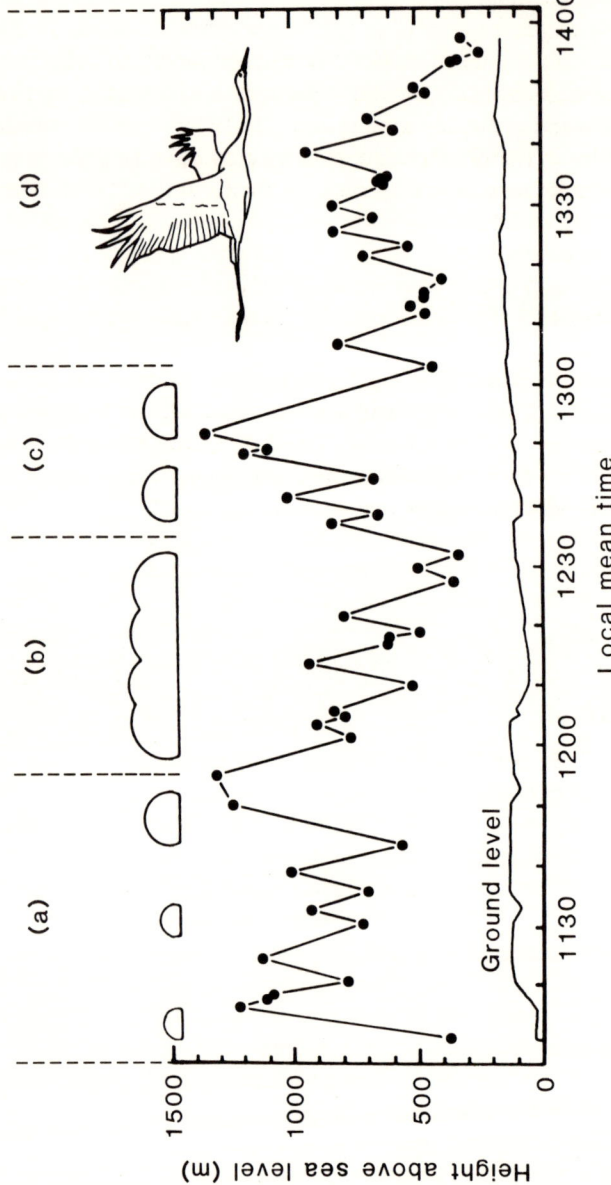

Figure 3.5 Record of height at the start and finish of each climb in a thermal versus time for a flock of migrating cranes, *Grus grus*, tracked by a light aircraft. During (*a*) there was a sunny sky, there were small cumulus clouds and thermals were good. During (*b*) the sky was overcast, the thermals weaker, and the birds gradually lost height. During (*c*) the thermals were good and the birds gained height. During (*d*) there were weak, inconsistent thermals and the cranes lost height. Eventually, when approximately 150 m above the ground, they began to flap their wings. (After Alerstam, 1981.)

3.3.4 *Thermals and updraughts*

Assistance from meteorological factors is not confined to tailwinds. Upward movement of air can result if there is an unstable, vertical distribution of temperature in the atmosphere. Birds can gain lift in these thermals by circling in their centre and then leaving them to glide in the migratory direction. If thermals are inadequate, the birds resort to flapping flight (Figure 3.5). Many birds, such as storks and eagles, migrate between Europe and the southern Sahara mainly by thermal soaring. However, thermal soaring is not possible over the sea so these birds follow a route which keeps them overland, but extends the distance travelled by approximately 1500 km.

Slope soaring (see section 2.7), using ocean waves as slopes, is the main method of locomotion in most small and medium-sized petrels. Many birds of prey take advantage of slope lift over a land slope when searching for food, but long-distance overland movement using slope soaring is unlikely because of the poor distribution of the slopes (Pennycuick, 1975).

3.3.5 *Formation flight*

Flying in flocks in V-formation or even in line can also be energetically advantageous, particularly for those birds, such as ducks and geese, with short wings. Birds flying in the upwash of the vortices created by the birds in front benefit by large reductions in profile and induced power (see Figure 2.3).

CHAPTER FOUR

THERMOREGULATION

Although migration enables birds to escape the winter extreme of their breeding grounds, those that breed in the far north are exposed to lower temperatures than they would be if they remained in their temperate feeding grounds. Even in hotter climates, temperatures can be very low at night and some birds actually remain in polar and sub-polar regions so the ability to conserve heat and maintain body temperature when exposed to an extremely low environmental temperature is a necessity for most birds. Except in a few hot deserts at midday or when standing exposed to the sun, a bird is unlikely to be in a situation where there is a net inflow of heat from the environment (Taylor, 1977). It is a strange fact, indeed, that high solar radiation on clear summer days in the Antarctic may impose a sufficiently high heat load on brooding penguins to cause them to pant. Overheating, or hyperthermia, usually results from the inability to remove metabolically produced heat quickly enough to the environment, either as a result of a reduced thermal gradient between the bird and the surrounding air and/or an increase in the rate of heat production.

4.1 'Normal' body temperature

Generally speaking birds, like mammals, do maintain deep body or 'core' temperature (T_c) within a very narrow temperature range. Although core temperature may appear to be one of the simplest and easiest of physiological variables, to measure, it is not easy to obtain a 'standard' value. Birds of different sizes have different thermal capacities and mass-specific metabolic rates (metabolic rate per unit mass). Handling small birds, for the introduction of a cloacal temperature probe, can cause a rapid increase in T_c above the prehandling level. There is even evidence to indicate that cloacal temperature measured by a probe may be approximately 0.5°C lower than intraperitoneal temperature measured at the same time by an indwelling radiotransmitter. To these sources of variation must be added diurnal and maybe seasonal changes. For example deep body temperature during the active phase of the daily activity cycle (day or night) can be 1°–3°C higher than that during the inactive phase in normothermic (i.e. non-torpid) birds. Care must be taken, therefore, to obtain T_c from birds under reasonably standard conditions.

Many species of birds, when 'inactive at moderate ambient temperatures in the waking phase of their daily cycle' have deep body temperatures between 39° and 43°C (Dawson and Hudson, 1970). However, these authors also point out that a number of species representing the Orders Sphenisciformes (penguins), Struthioniformes (ostriches), Casuariiformes (Emu and cassowaries), Apterygiformes (kiwis), Gaviiformes (loons), Podicipediformes (grebes) and Procellariiformes (albatrosses, shearwaters, petrels etc.) have deep body temperatures in the range 37°–41°C. Of these, the first four Orders consist entirely of flightless birds in which body temperature is below 40°C and within the range of that of eutherian mammals. The other three Orders contain aquatic birds, which in itself cannot be significant, because other aquatic birds have higher deep body temperatures.

Regardless of its actual value, if deep body temperature is substantially above ambient temperature and dependent upon internal heat generation, as it is in birds, then the expenditure of substantial amounts of energy is involved in the maintenance of that temperature. Even though few birds appear to hibernate as such, the daily reduction of deep body temperature (torpor) is used to conserve energy by some birds.

4.2 Torpor

Goat suckers (Caprimulgidae) and hummingbirds (Trochilidae) exhibit well-developed patterns of torpor. Some species of swifts may also become torpid on occasion. The Poorwill and other caprimulgids can enter a state of torpor lasting several days if the food supply is inadequate to support normal body temperature. This has been termed hibernation. As ambient temperature falls at night (or in the laboratory), these birds may become torpid. Oxygen uptake, deep body temperature (Figure 4.1), heart rate and respiratory frequency all decline. During entry into torpor, not only does heat production fall but heat loss may be enhanced, as indicated by an increase in thermal conductance, presumably as a result of peripheral vasodilatation, exposure of the extremities and a reduction of the air layer trapped in the feathers. In the laboratory, body temperature can become

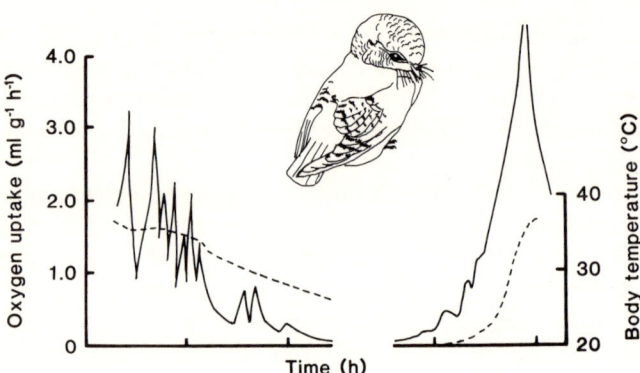

Figure 4.1 Oxygen uptake (———) and body temperature (---) of a poorwill, *Phalaenoptilus nuttalli*, during a typical torpor cycle. (After Withers, 1977.)

too low to allow spontaneous arousal (Calder and King, 1974), but it appears that in three species of hummingbird at least, body temperature is actively regulated at a lower level during torpor. Two highland species regulate at 10°–12°C, whereas a lowland species regulates at 18°–20°C. These compare with average minimum environmental temperatures of 4°–6°C and 16°–19°C respectively. In other words, oxygen uptake and body temperature decline with environmental temperature until a critical value is reached, when oxygen uptake increases and body temperature remains constant or even increases slightly (Figure 4.2). Like hibernating ground squirrels the birds are still homeothermic, but the thermostat is at a lower 'set point'.

Torpidity does not necessarily occur every night (when the birds are not feeding), and Lasiewski (1963) suggested that it may only occur if the energy reserves of the bird are low. For one species of hummingbird, there is evidence to suggest that whether or not the bird enters torpor does indeed depend on the energy reserves reaching a certain (low) threshold level. If this threshold is reached, perhaps as a result of insufficient food intake during the day, then energy expenditure is greatly reduced by a fall in body temperature. Important though this is, the fact that it does not occur regularly may imply that there are risks involved, e.g. from predators or the inability to arouse.

Arousal may be triggered by an internal rhythm or by external stimuli such as light, noise etc. (Lasiewski, 1963). Shivering thermogenesis (see section 4.4.1) and reduced thermal conductance (peripheral vasoconstriction, ptiloerection, i.e., erection of the feathers and covering of the extremities) ensure a rapid increase in body temperature. It is important that

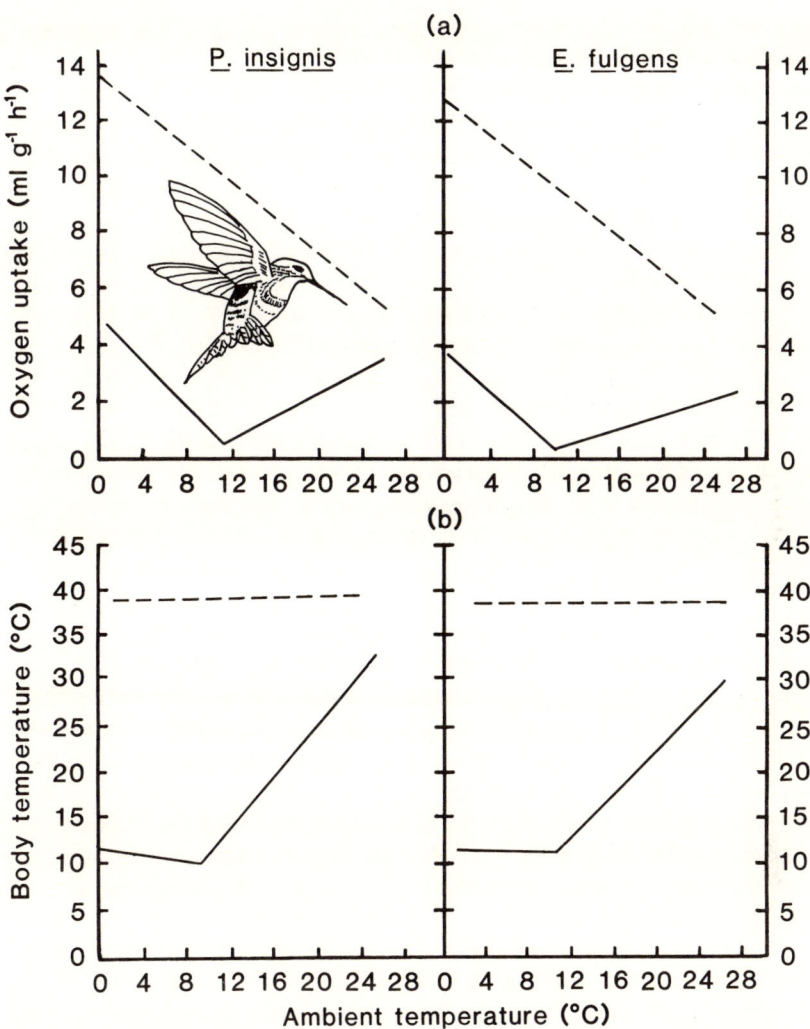

Figure 4.2 Relationship between (a) oxygen consumption and ambient temperature (b) body temperature and ambient temperature in the hummingbirds *Panterpe insignis* and *Eugenes fulgens* while at rest (---) and during torpor (——). (After Wolf and Hainsworth, 1972.)

both entry into torpor and arousal occur as quickly as possible. Because of their higher surface area/volume ratios, smaller birds can cool faster than larger ones, and because of their higher mass-specific metabolic rate and smaller mass, they can also warm up more quickly. The ratio of metabolic rate during torpor to that in resting birds at normal body temperature is inversely related to body mass, being approximately 1:55 in hummingbirds and 1:13 in the larger Common Night Hawk. However much energy is saved during torpor, some extra energy is used in the arousal process, but only 1/85 of the daily energy expenditure of a 4g hummingbird is used in warming it from 10° to 40°C (Calder and King, 1974). Thus, intermittent, short periods of torpor are very economical for small birds. Their size makes it both necessary and possible.

Birds of intermediate size, such as the Poorwill, may remain torpid for several days, and even larger birds may reduce body temperature, but by smaller amounts. The desert-dwelling Roadrunner and the Turkey Vulture can both conserve energy by reducing deep body temperature by 3°–4°C at low ambient temperatures and by sunning themselves (heliothermia) during the day. The combined use of hypothermia and heliothermia could produce energy savings of up to 40% in the insectivorous Roadrunner, which would be particularly important during the winter months when there is a decline in insect numbers.

4.3 Thermoneutral range

When a bird is at rest and in a post-absorptive state, the metabolic rate (standard metabolic rate) is sufficient to maintain a constant deep body temperature over a range of ambient temperature—the thermoneutral range. At one end is the lower critical temperature, at the other end the higher critical temperature. Below and above these temperatures heat production (oxygen uptake) increases. It can be seen from Figure 4.2 that oxygen uptake increases in non-torpid hummingbirds as environmental temperature falls below 28°C. For most birds the thermoneutral range extends to much lower temperatures. Strangely, perhaps, it is not clearly related to body size (i.e. higher for smaller birds), but more to natural environmental temperatures (Calder and King, 1974). For example, in the Snow Bunting the lower critical temperature is 9°C, whereas in the Common Cardinal it is 18°C. Both have a mass of approximately 40 g. In the aquatic dipper (50 g) which inhabits cold mountain streams, lower critical temperature is 11.5°C, whereas in the non-aquatic Evening Grosbeak (55 g), it is 16°C.

4.4 Cold exposure

Although standard metabolic rate is sufficient to maintain body temperature over the thermoneutral range, it must not be assumed that no

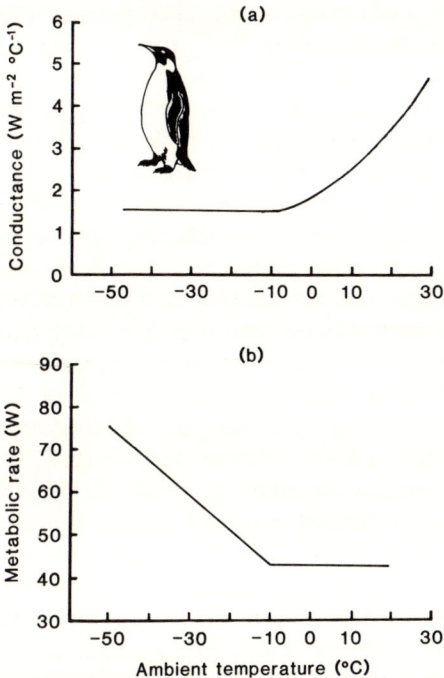

Figure 4.3 Relationship between (a) thermal conductance and environmental temperature (b) oxygen uptake and environmental temperature in Emperor Penguins, *Aptenodytes forsteri*. (After Pinshow *et al.*, 1976.)

regulation is occurring. What it does mean is that the changes in the insulative properties of the bird are sufficient to maintain body temperature. Thermal conductance decreases as ambient temperature falls from the top end to the bottom end of the thermoneutral range (Figure 4.3). Only when these mechanisms are inadequate to prevent heat loss exceeding heat production, does heat production need to increase.

4.4.1 *Shivering thermogenesis*

Shivering is thought to be the only form of extra heat production in birds (i.e. in excess of standard metabolism and specific dynamic action of food), although Oliphant (1983) has found fat deposits in Ruffed Grouse and Black-capped Chickadees with histological features typical of mammalian brown adipose tissue. The physiological significance of this is unknown.

There is certainly a direct relationship between electrical activity in the pectoral muscles of birds and reduced environmental temperature below the lower critical temperature. Also, the increased oxygen uptake during such exposure can be abolished in pigeons by preventing muscle activity (Hart, 1962). Thus there is no physiological evidence to suggest the use of non-shivering thermogenesis in birds involving brown adipose tissue as in mammals. It is the fast glycolytic fibres (type IIb) in the pectoral muscles of pigeons that are responsible for brief but vigorous periods of shivering thermogenesis.

It has been assumed that during exercise, shivering is suppressed and the extra heat produced is sufficient to maintain body temperature at low environmental temperature (see, for example, Berger *et al.*, 1971). If this is so, steady-state oxygen uptake at a given level of activity should be independent of environmental temperature. This is certainly the case for Budgerigars flying at ambient temperatures from 37° to 18°C and for Japanese Quail running at ambient temperatures from 30° to 10°C, particularly at higher running speeds. For two species of hummingbird, however, this is not so. There is a linear relationship between oxygen consumption during hovering and ambient temperature (Figure 4.4). This is clearly an area that warrants further study, but it is perhaps worth noting that when foraging in water at 2°C, energy consumption of King Penguins may be 74% of that during periods of inactivity in water at the same

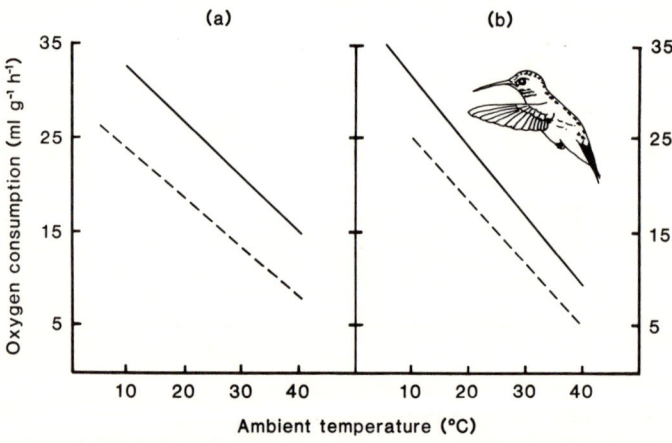

Figure 4.4 Relationship between oxygen consumption and ambient temperature in two species of hummingbird, (*a*) *Amazilia cyanifrons* and (*b*) *Amazilia tzacatl*, while at rest (—) and while hovering (———). (After Schuchmann, 1979.)

temperature (Butler and Jones, 1982). Perhaps different birds respond differently.

Shivering thermogenesis is more effective in some winter- (i.e. cold)-acclimated birds than in summer ones (Hart, 1962). It is generally thought that improved cold resistance during winter, in small passerines at least, is the result of metabolic adaptations rather than of a significant increase in insulation. Although the ability to increase heat production is not vastly greater in winter birds, they are able to maintain these elevated levels for much longer periods than summer birds (Figure 4.5). Most species that spend the winter in cold regions have significantly greater fat reserves at the onset of roosting in winter than in summer, although they are usually smaller than those before migration (Dawson *et al.*, 1983). Fat stores during winter are not for long-term use and a large part of fat stored during the day is used at night. Work on American Goldfinches has led to the suggestion that the ability to sustain a higher rate of shivering thermogenesis for longer periods in winter results from a greater ability of the fast oxidative glycolytic fibres (type IIa) in the pectoralis muscles to use fatty acids. As during migration, this would conserve carbohydrate reserves thereby reducing the likelihood of fatigue. Unfortunately there is no biochemical evidence to support this hypothesis: no seasonal difference has been found

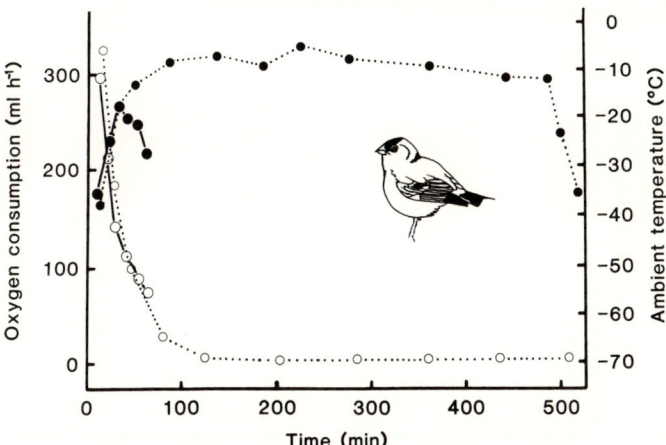

Figure 4.5 Response of oxygen consumption (●), as ambient temperature (○) is reduced to −70°C, in a winter-acclimatized (·····) and a summer-acclimatized (———) American Goldfinch, *Carduelis* (*Spinus*) *tristis*. Although oxygen uptake increases in the summer bird, it is not maintained for much more than an hour. In the winter bird the elevated oxygen uptake is maintained for 8 h. (After Dawson and Carey, 1976.)

in the capacity of the pectoralis muscle of goldfinches to oxidize fatty acids.

An important aspect of survival in winter for some birds such as the Willow Ptarmigan is a reduction of the lower critical temperature. This may result at least in part by improved insulation from feathers and fat. It is possible experimentally to change the shivering threshold of pigeons from 21° to 14°C by exposing them to a short photophase under cold conditions (Saarela and Vakkuri, 1982). The authors did not test the effect of a short photophase at higher temperatures, but they conclude that the pineal plays an important role in temperature regulation in birds.

4.4.2 Circulatory adjustments

In pigeons exposed to 2°C, delivery of oxygen to the shivering muscles is achieved by a 25% rise in cardiac output and a 40% increase in $(C_aO_2 - C_{\bar{v}}O_2)$. On the other hand, in Pekin ducks at $-20°C$, there is little change in $(C_aO_2 - C_{\bar{v}}O_2)$ but cardiac output is 2.5 times the value at an environmental temperature of 20°C. If oxygen supply to the muscles is insufficient, even at room temperature, deep body temperature falls in chickens, so it would be interesting to study thermoregulation in birds at high altitude where both oxygen availablity and environmental temperature are low.

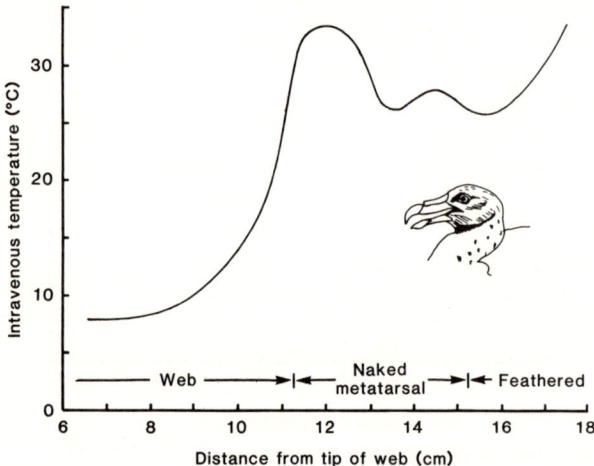

Figure 4.6 Temperature of venous blood along the lower part of the leg of a Giant Petrel, *Macronectes giganteus*, while the web was immersed in ice-cold water. (After Johansen and Millard, 1973.)

The circulatory system is also important in controlling heat loss from the body. As might be expected, heat loss through the extremities, such as the legs in gulls and herons, is greatly reduced in a cold environment—air or water. This may be achieved by a countercurrent arrangement of the blood vessels and/or by arterio-venous (A-V) anastomoses bypassing the surface vessels. A reduction in body temperature of the Giant Petrel causes a decline in blood flow to the foot. There is anatomical evidence of a countercurrent exchange system in the leg of the petrel and a marked increase in the temperature of venous blood on its way along the leg (Figure 4.6) is physiological evidence that effective heat exchange occurs between warm arterial blood from the body to the cold venous blood from the foot.

Sudden immersion of the foot into sea water at $-2°C$ causes a transient increase in blood flow ('cold flush') which is followed by a progressive fall in flow. The subsequent reduction in blood flow is the result of vasoconstriction and an increase in viscosity of the blood as it cools. After approximately 3 min in cold water, the vasoconstriction in the foot is periodically interrupted by transient increases in flow ('huntings'). The significance of the initial increase in blood flow upon cold immersion is unclear. It may serve to prevent cold shock to the peripheral tissues with impairment of sensory and motor functions. In the petrel it is nervously mediated, although the precise mechanism is unclear. According to Johansen and Millard (1974) it is, at least in part, the result of a cholinergic dilator mechanism, whereas according to Murrish and Guard (1977) it is entirely the result of dilatation mediated by β-adrenergic receptors.

It is interesting to note that in mammals nerves from heterothermic tissues, such as the leg, conduct impulses at lower temperatures than those from warmer regions of the body. Nonetheless, there is evidence that at very low foot temperatures (8°C in the domestic duck) the vasoconstrictor response to nerve stimulation and to low doses of adrenalin and noradrenalin is negligible. This would allow a brief period of vasodilatation and increased blood flow until the tissue rewarms and vasoconstriction occurs again, and such a mechanism would certainly explain the 'hunting reaction' seen in petrels. On the other hand, Murrish and Guard (1977) conclude that following the initial 'cold flush' upon exposure to the cold, blood flow to the foot in Giant Petrels is reduced by vasoconstriction of arterio-venous anastomoses and of the countercurrent exchange vessels. This vasoconstriction is mediated via α-adrenergic receptors. Viscosity of the blood increases 2.7 times as it cools from 30° to 0°C, so this also causes a reduction in peripheral blood flow. To prevent the feet from freezing at very low temperatures the A-V anastomoses periodically dilate, as a result of stimulation of β-adrenergic receptors, allowing warm blood temporarily to bypass the heat exchanger and enter the feet. According to this scheme, the countercurrent heat exchanger is under dual sympathetic control and the 'hunting reaction' is entirely an active process.

Another interesting example of superficial blood vessels dilating when exposed to something cold are those of the brood patch of incubating hens. If the patch is in contact with cool eggs, there is an immediate increase in blood flow which appears not to be nervously controlled; it is purely a local response. It is not certain whether this vasodilatory mechanism in response to a cold stimulus is always present in breast skin or whether it is restricted to the new blood vessels that develop in the brood patch. A permanent response of this type would be disadvantageous during cold weather. It is known that full brood-patch vascularization can be elicited in the California Quail by administration of oestrogen and prolactin, and it is tempting to suggest that one or both of these hormones could influence the sensitivity of the blood vessels in the brood patch to thermal stimuli, causing them to dilate when cooled.

4.4.3 *Respiratory adjustments*

The increased demand for oxygen during cold exposure is met in pigeons exposed to 2°C by a 50% increase in respiratory tidal volume and a 30% increase in respiratory frequency. There is no change in the effectiveness of gas exchange, as the fraction of oxygen removed from the inspired gas (O_2Ext) remains constant. In other words the proportional rise in oxygen consumption is matched by an equivalent proportional increase in ventilation. This appears not to be the case in a number of other birds that have been studied.

In the Kittiwake, European Coot, Parrot and Pekin duck oxygen uptake increases by a greater proportion than ventilation, which means that O_2Ext increases. An extreme example is the coot in which at -25°C ventilation is unchanged from that at 25°C, whereas oxygen uptake is 3.4 times greater. Parabronchial oxygen extraction increases from 0.27 at 25°C to 0.62 at -25°C (Brent *et al.*, 1984). Precisely how this increase in the effectiveness of gas exchange is achieved is not known but the physiological significance is that heat loss from the air ventilating the lung is lower than it would be if ventilation increased by the same proportion as oxygen uptake. This heat conservation does not appear to occur at the expense of oxygenation of the blood, for in ducks arterial P_{O_2} is exactly the same in animals exposed to -20°C as it is in those at 20°C (Bech *et al.*, 1984).

4.4.4 Penguins in the Antarctic

Studies on penguins may indicate the kinds of adaptations necessary for survival in a continually cold environment. They have a wide thermoneutral range. In Humboldt Penguins it is from 2°–30°C and in the larger and more southerly-dwelling Emperor Penguin, it is from $-10°$ to 20°C (Pinshow *et al.*, 1976). As the latter authors point out, however, the lower critical temperature is not particularly low, considering that air temperatures of $-30°$ to $-40°$C are experienced at the rookeries and the thermal conductance of penguins is within the range of that for other birds. They do have a slightly smaller surface area than would be expected for birds of similar mass. There are anatomical arrangements of the arteries and veins in the legs and flippers of penguins which suggest that countercurrent heat exchange occurs. Of particular interest in this context is the post-orbital *rete mirabile* (usually called the *rete mirabile ophthalmicum*), which is extremely well adapted to prevent excessive heat loss from poorly insulated regions of the head. Body temperature of Adélie and Emperor Penguins (37.8°–38.9°C) is at the lower end of the range for other birds, but standard oxygen uptake is no more than 27% higher than the value predicted for birds of similar mass from allometric formulae (Butler and Jones, 1982). Penguins, like other birds, are able to conserve water and heat when breathing in cold air by what has been called a temporal countercurrent heat exchanger in the nasal passages (Murrish, 1973). Inhaled air is warmed and saturated on its way to the lungs and at the same time cools the mucosa lining the nasal passages. Exhaled air is therefore cooled as it passes over the mucosa and water condenses out. When in air at 5°C, Adélie and Gentoo Penguins cool expired air to 9°C, thus saving 82% of the water and 83% of the heat added to the inhaled air. This is not a passive system. Blood flow through the mucosa is controlled by an α-adrenergic mechanism. There do not, therefore, appear to be any particular physiological characteristics for surviving in the cold that distinguish penguins from other birds. A continuous supply of energy (food) must be of vital importance.

Although both Adélie and Emperor Penguins may walk long distances to their rookeries, emperors do so in the Antarctic winter when, as already mentioned, air temperature can be as low as $-30°$ to $-40°$C and wind velocity as high as 40 m s^{-1}. Having laid the egg, the female returns to sea to feed but the male stays for an average of 64 days to incubate the egg. A single isolated male would require approximately 25 kg of fat to fast in the prevailing conditions for 100 days (i.e. including the time to walk to and from the rookery). Add this to the 1.5 kg of fat required to walk the distance (see section 2.9.1) and the total easily exceeds the actual fat reserves of 15–20 kg in a large male. The secret of survival is that the male penguins huddle together, thereby reducing the area of the body exposed to the cold air (Pinshow *et al.*, 1976). Those at the outside of the group periodically move to the inside. While inside the huddle, the penguins may metabolize at or below their resting rate, but if outside the group they can be seen to shiver, which is associated with increased metabolic rate. When in a huddle, energy expenditure may be further reduced by the penguins becoming hypothermic. Huddling and other forms of behavioural thermoregulation have been described for other birds during winter, e.g. Black Duck (Brodsky and Weatherhead, 1984).

Emperor Penguins also shiver when at rest in Antarctic water, and it has been shown for the Adélie Penguin that when inactive in water at $-1.85°C$ oxygen uptake is approximately four times the resting value in air at $13°–15°C$. Increasing the depth of submersion reduces the air layer in the feathers, causing a rise in thermal conductance and an even greater increase in oxygen consumption. Presumably even the larger Emperor Penguins are affected in a similar way by immersion into water. It is assumed that penguins are active almost all of the time they are in the water and that they get on to ice or land frequently to preen and condition their feathers. Waterproofing must be vitally important to them.

4.4.5 *Nervous mechanisms*

The area of the CNS which is particularly important in the thermoregulatory responses to cold is located in the spinal cord in most birds and not in the preoptic and anterior hypothalamic region of the brain stem (POAH), as in mammals. Cooling of the hypothalamus causes little or no increase in shivering (and thus in oxygen uptake) in pigeons, Adélie Penguins, California Quail or ducks. In fact, in ducks and penguins cooling the hypothalamus inhibits shivering and activates heat dissipation. It has been suggested that there is an effect on ptiloerection and blood flow to the skin in pigeons during cooling of the hypothalamus, although direct measurements indicate that there is only a weak effect on the cardiovascular system in ducks (Bech *et al.*, 1982). In contrast, cooling of the spinal cord does induce shivering and reduced peripheral blood flow in all the birds studied, although in the Adélie Penguin the metabolic response is apparent only at low environmental temperatures of approximately $-20°C$. In ducks there is a region in the brain stem in the lower midbrain/upper pontine area which appears to be sensitive to cooling and to elicit clear but weak metabolic and circulatory responses. In the whole animal this response may actually be counteracted by an opposite response originating from the anterior hypothalamus where, unlike the situation in mammals, the Q_{10} of synaptic transmission of the cold signal input is greater than that of the warm signal input (Lin and Simon, 1982). This could explain the paradoxical situation where cooling of the hypothalamus causes a decrease in metabolic heat production and increased heat loss so that body temperature falls. Of the birds studied so far, the emu is an exception in that cooling the hypothalamus from 38° to 34°C does cause shivering, and only if hypothalamic temperature reaches 30°C is the 'paradoxical' response of heat dissipation apparent.

It appears that thermoreceptors in the CNS (brain stem, cervical and thoracic regions of the spinal cord) and skin represent a small fraction (25%) of the total cold sensitivity of the body in ducks (Inomoto and Simon, 1981). It is concluded by these authors, therefore, that most of the cold sensitivity in these, and possibly in all, birds depends on deep body temperature receptors outside the CNS. These receptors appear to be more important in birds than in mammals and this may result from the weak hypothermic response of the hypothalamus in birds.

4.5 Heat stress

Heat can be lost by radiation, conduction and convection (dry heat loss) and by evaporation, when water as well as heat is lost. In a resting bird there is no reduction in heat production below the standard rate, but during exercise heat production actually increases.

4.5.1 *Dry heat loss*

Dry heat loss is enhanced by increasing blood flow to the skin, particularly the legs and feet, and by keeping these structures (and the whole body if possible) in shade and in a flow of air. During flapping flight, Herring Gulls can lose up to 80% of total heat production through their feet (Baudinette *et al.*, 1976), although the authors do point out that this process may not be so effective in birds without webbing between the digits. Despite the effectiveness of the feet, deep body temperature does increase, by approximately 1.5°C, during flapping flight in the gulls. Compressing the contour feathers and holding the wings away from the body so as to expose the thinly feathered sides of the thorax and underside of the wings are also important behavioural adjustments to heat stress. During high external heat loads, and often when active, dry heat loss may be insufficient to maintain body temperature at its normal value. Birds may then store heat and allow body temperature to rise or become dehydrated in the process of evaporative heat loss.

4.5.2 Heat storage

There is no doubt that a number of birds allow body temperature to rise by several degrees centigrade during heat stress resulting from high external temperatures (Dawson and Hudson, 1970) and during exercise in air (Torre-Bueno, 1976). The stored heat can then be dissipated during cooler conditions, e.g. at night, or at the end of the period of exercise. During exercise, heat storage can amount to a sizeable fraction of heat production, depending in part on the mass of the bird, the increase in body temperature and the duration of the exercise. In Rhea (18–25 kg) running at $10\,\mathrm{km\,h^{-1}}$ for 20 min at an environmental temperature of 25°C, approximately 30% of total heat production is stored, whereas at 43°C, 75% of heat is stored. The smaller White-necked Raven (0.5 kg) stores 10% of total heat production when flying at $10\,\mathrm{m\,s^{-1}}$ for 20 min at an ambient temperature of 28°C. However, once a steady-state temperature has been reached, that is the limit of heat storage, and a flying Starling produces enough heat to raise its body temperature by 4°C in 2 min. Torre-Bueno (1976), in fact, found that body temperature increases even in starlings flying at 0°C. He concluded that birds adjust their insulation to allow a certain increase in body temperature and proposed that the functional significance of hyperthermia during flapping flight is related to increases in maximal work output and muscle efficiency. It is possible, therefore, that elevation of body temperature may serve different functions under different conditions.

4.5.3 Brain temperature

Whatever the functional significance of the increase in deep body temperature may be, brain temperature is maintained at a lower level. In fact, even in birds under resting, thermoneutral conditions, there is approximately a 1°C difference between brain temperature and body temperature in mallard ducks, pigeons, kestrels and rheas. This difference is at least maintained as the birds become hyperthermic during increasing external

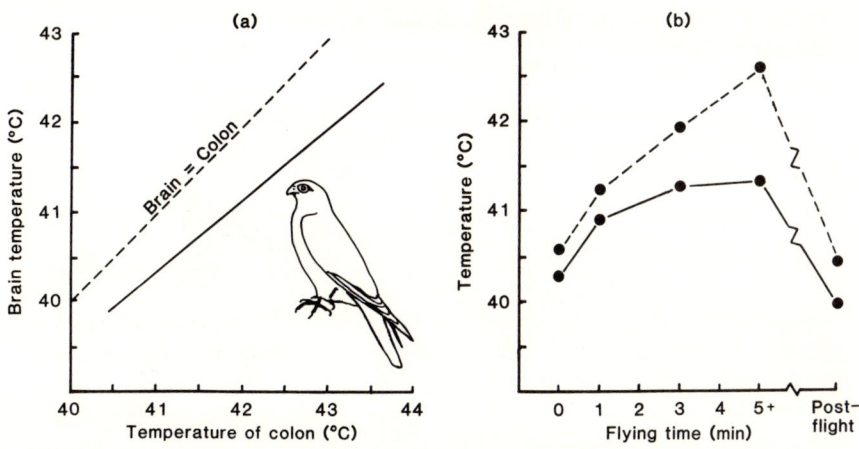

Figure 4.7 (a) Relationship between colonic temperature and brain temperature (———) in resting kestrels, *Falco sparverius*, as environmental temperature increases between 22.5° to 36.1°C. Dashed line represents isothermal points. (b) Brain and colonic temperatures in kestrels before, during and after flapping flight at 10 m s^{-1} with an ambient temperature of 23°C. Data at 5 + min obtained between 5 and 15 min after onset of flight and represent steady-state values. (After Bernstein *et al.*, 1979.)

heat loads and during exercise, and in some cases it may even become greater (Figure 4.7). The structure which appears to be important in maintaining the brain cooler than the body in birds is the *rete mirabile ophthalmicum*, RMO (cf. penguins where its function is probably to prevent heat loss). It has been found in all birds that have been examined, although it is poorly developed in the Zebra Finch, and this bird is unable to cool its brain below the temperature of the body. The *rete* is closely associated with the circulatory system of the eye, hence its name, and warm arterial blood from the body is cooled by the counter-current exchange with venous blood returning from the relatively cool beak, evaporative surfaces of the upper respiratory tract and the eye (Figure 4.8).

The importance of perfusing the RMO during heat stress may be the explanation for the 3 × increase in carotid blood flow during thermal panting in domestic ducks while ischiadic blood flow does not change. More direct evidence for its importance can be obtained by interrupting blood supply to the ophthalmic rete of pigeons. Brain temperature then becomes 0.4°C *higher* than body temperature during heat stress. A similar effect can also be achieved in heat-stressed pigeons by preventing evaporative heat loss in the head. With the birds breathing through a tracheal cannula, the eyes covered, the nares plugged and the mouth taped closed,

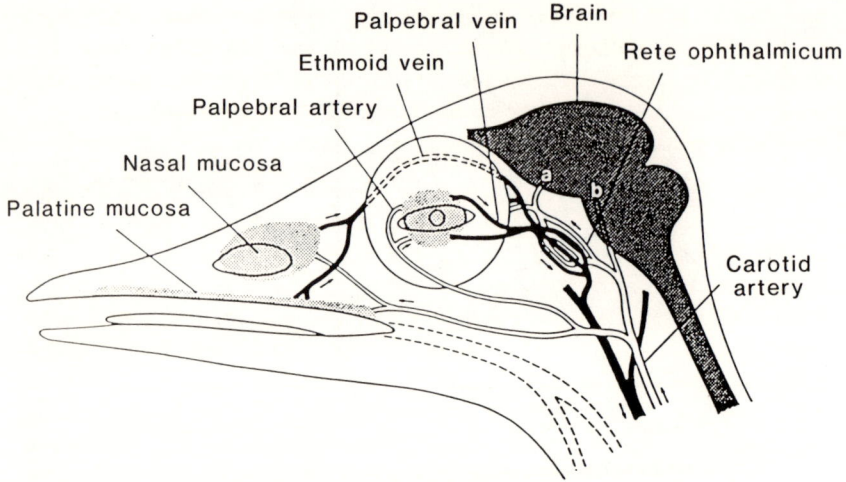

Figure 4.8 Diagram showing the relationship of the *rete ophthalmicum* to the eye and the brain. (*a*) indicates the anterior blood supply to the brain, via the rete, and (*b*) shows the posterior route via the cerebral carotid artery. Brain temperature may be regulated by mixing of blood from the two pathways. The areas that are important in heat dissipation are shaded — nasal and buccal cavities, eyelids and cornea. (Midtgard, 1983.)

brain temperature exceeds body temperature. The role of the eyes under these conditions is not obvious. Plugging the nostrils and taping the mouth (with the eyes uncovered) causes a significant lowering of the body/brain temperature difference, but not a reversal, whereas covering the eyes alone has no significant effect on body/brain temperature difference. Corneal cooling is probably more important during flight when there is a flow of air over the eyes. There seems no doubt though that cooling the surfaces in the upper respiratory passages plays an important role in maintaining a relatively low brain temperature. It is possible that the major function of respiratory evaporative heat loss during panting and gular flutter is related to this role.

4.5.4 *Evaporative heat loss*

Evaporative heat loss becomes a more important component of the thermoregulatory process as the need to dissipate heat exceeds dry heat loss. Because water is lost, it can present problems if this loss (together with urine loss etc.) exceeds metabolic water production and little or no drinking water is available (hence heat storage and behavioural mechanisms to improve dry heat loss). The use of the respiratory system involves enhanced ventilation and therefore greater energy expenditure (more heat production !), so as environmental temperature increases there is a point where oxygen uptake rises above the standard level. This denotes the upper end of the thermoneutral range and is called the upper critical temperature.

(a) *Cutaneous evaporation.* Despite the absence of sweat glands and the covering of the body in contour feathers, evaporative heat loss does occur across the skin in most birds that have been studied. A notable exception is the ostrich. At an ambient temperature of 40°C, total evaporative water loss from a 90kg animal is approximately 3.5 g min^{-1} and less than 2% of this is across the skin. At the other extreme a 12.5 g Zebra Finch at an ambient temperature of 30°C loses $1.8 \text{ mg } H_2O \text{ min}^{-1}$ and 63% is across the skin. Thus birds can lose heat and water across their skins. Skin permeability and plumage thickness are two important factors influencing cutaneous water loss, and cutaneous evaporation decreases substantially during the first 4 weeks after hatching in the Painted Quail. Another important factor is the difference in water vapour pressure between the evaporative surface and the surrounding air, and this can be affected by air circulation or ventilation. Such convection will remove humid air and maintain a greater vapour pressure difference across a thinner layer of boundary air. Panting and gular flutter cause the necessary convection across the respiratory surfaces.

(b) *Respiratory evaporation (panting).* Panting threshold is reached at body temperatures between 41°–44°C and the onset of panting in resting birds may be an abrupt or gradual increase from the normal pattern of ventilation. During panting, evaporation occurs from the nasal, buccal and upper pharyngeal surfaces as well as from the trachea. Evaporative cooling may be augmented by gular flutter which involves the hyoid apparatus and associated musculature moving the gular region. Under these conditions evaporation occurs from the moist surfaces of the pharynx and anterior oesophagus as well as from those of the buccal cavity. It has been suggested that gular flutter is energetically less costly than panting (Calder and King, 1974). Pigeons pant at a frequency which is close to the resonant frequency of the respiratory apparatus (see section 2.6.1). The fact that panting and flutter frequencies of certain species cover a narrow range and are relatively independent of heat load suggests that resonant oscillation, and hence minimum energy expenditure, may occur in a number of birds (Dawson, 1982). If birds are already dehydrated when exposed to thermal stress, then evaporative water loss is not as great as normal and body temperature increases more than usual.

As well as excessive water loss, which is unavoidably associated with panting and gular flutter, there is also the problem during panting of excessive removal of CO_2 causing hypocapnia and alkalosis. There is no doubt that ventilation exceeds metabolic demands during heat stress in both resting and exercising birds, and this usually occurs at a lower ambient temperature in the latter case (Figure 4.9). However, hypocapnia and alkalosis are not always as great as expected. During moderate heat loads, arterial P_{CO_2} decreases by only 0.3–0.4 kPa (and pH increases little) in Bedouin Fowl, ducks, Mute Swans, pigeons and Adélie, Chinstrap and Gentoo Penguins, despite the hyperventilation. This is because, associated with the large increase in respiratory frequency (20–30 times resting), there is a large reduction in tidal volume (to 15–50% resting) (see Dawson, 1982), so that effective ventilation of the gas exchange surfaces increases very little. The Mute Swan decreases tidal volume so that it is only slightly greater than dead space ventilation. Pigeons employ what has been termed 'compound ventilation' during which a high-frequency, low-volume component is superimposed on a low-frequency, large-volume component

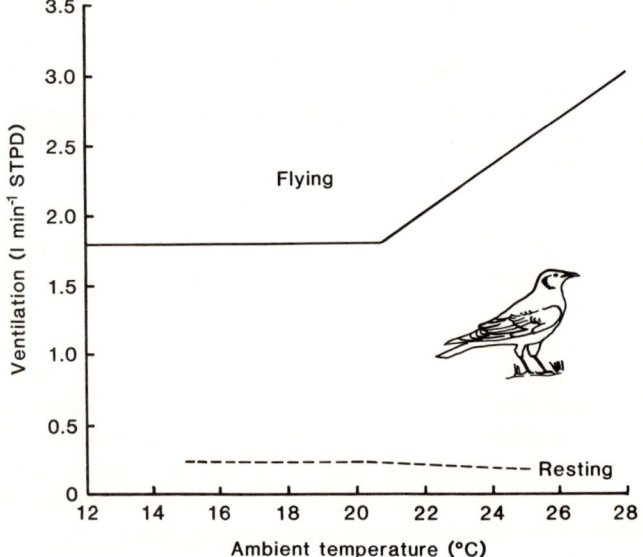

Figure 4.9 Relationship between minute ventilation volume and ambient temperature in the Fish Crow, *Corvus ossifragus*, while at rest (---) and during steady-state horizontal flapping flight (——). Note that overall hyperventilation during flight occurs at lower ambient temperature than it does in resting crows. (After Bernstein, 1976.)

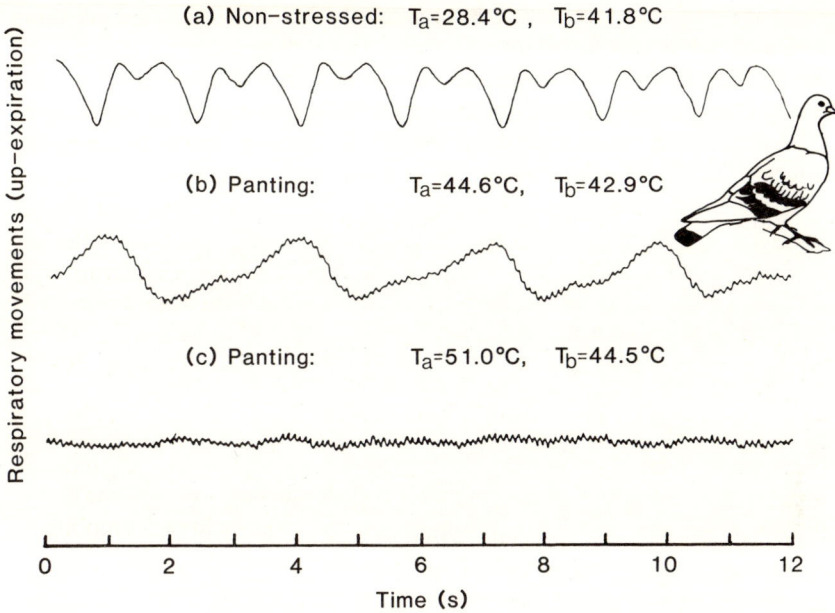

Figure 4.10 Respiratory patterns in resting pigeons at different ambient (T_a) and body (T_b) temperatures. (*a*) Normal ventilation in absence of heat stress. (*b*) Compound ventilation under mild heat stress, where rapid low-amplitude component is superimposed on slow high-amplitude component resembling the normal pattern. (*c*) Simple panting under severe heat stress. Tidal volume is approximately 3 times greater than that of the fast component during compound ventilation. (After Ramirez and Bernstein, 1976.)

resembling the normal pattern (Figure 4.10). This type of ventilation may also be seen in White-necked Ravens when flying. The Greater Flamingo uses a different pattern still. During panting tidal volume is slightly less than dead space, but periodically there are short bursts (1–3 breaths) of increased tidal volume. At more extreme levels of heat stress (i.e. when body temperature is at 46°C in pigeons), pulmonary ventilation increases and a clear hypocapnic alkalosis occurs in most birds that have been studied. In pigeons the slow component of ventilation disappears (Figure 4.10) and the high-frequency component has a tidal volume closer to dead space volume. Such levels of heat stress are thought to be uncommon under natural conditions and birds are usually capable of coping with heat loads without becoming too alkalotic.

Even during severe heat stress, there is no reduction of arterial P_{CO_2} in the ostrich, which suggests that the increased air flow through the lung completely bypasses the gas exchange region. The relationship between air flow and blood perfusion through the lung may change quite dramatically during thermal stress in these large cursorial birds. For example, in the Emu, there is not only a slight reduction in arterial P_{CO_2} indicating *hyper*ventilation of the gas exchange surfaces, there is also a reduction in arterial P_{O_2} which is indicative of *hypo*ventilation (Jones *et al.*, 1983).

(*c*) *Cutaneous vascular plexus of Columbidae.* When ambient temperature reaches a certain level and the bird begins to absorb heat from the environment, evaporative heat loss may exceed total metabolic heat production (see Dawson, 1982). In rhea, most of this is via the respiratory route, but in pigeons almost 60% of total evaporative water loss is across the skin at high ($>50°C$) environmental temperatures and on occasions they may actually stop panting (Marder, 1983). This remarkable capacity for cutaneous evaporative cooling in pigeons is greater than in the partridge, and pigeons are more tolerant of high ambient temperatures than partridges. This may relate to the presence of a vascular plexus in the neck of Columbidae (doves and pigeons) which is cooled by air entering the oesophagus (Gaunt, 1980). This system has the advantage of convective air flow but does not carry with it the possible disadvantage of respiratory alkalosis.

4.5.5 *Flapping flight*

Heat loss during flapping flight is primarily by non-evaporative means except at high ($>30°C$) ambient temperatures, when it is approximately 50% of the total. While non-evaporative heat flow across the feet, legs and beak is no doubt important (some birds extend their legs during flight at high ambient temperatures), heat flow across the general body surface also contributes to dry heat loss. Changes in feather geometry, as well as in cutaneous blood flow, probably allow some control over this route. Despite the dominance of dry heat loss during flapping flight at relatively low ambient temperatures, evaporative water loss exceeds metabolic water production at ambient temperatures above 7°C for Starlings and at temperatures above $-10°C$ for the Black Duck. It has been suggested that during long migrations birds fly at an altitude where air temperature is sufficiently low for them to remain in water balance by relying almost exclusively on non-evaporative heat loss. Of course, birds, such as the White-necked Raven, which routinely flies when ambient temperatures at ground level are above 30°C, could reduce heat production during flight by soaring and gliding. Thus, all the strategies used by flying birds to keep energy costs to a minimum will also serve the important function of conserving water. At the end of a flying period, birds invariably pant, presumably to remove the heat accumulated during the exercise.

Although oxidation of a gram of fat produces almost twice as much water as oxidation of a gram of carbohydrate (1 g compared with 0.55 g respectively), the preferential use of fat during flight does not lead to a greater production of metabolic water. As already stated (section 3.3.1), more energy is produced from a gram of fat than from a gram of carbohydrate, so for a given metabolic rate the amount of water produced is similar for either substrate (approximately 0.4 g H_2O kJ^{-1}).

4.5.6 Panting and respiratory gases

Despite the fact that arterial P_{CO_2} may not fall as much as might be expected as a result of the increase in ventilation during hyperthermia, it does fall by a small amount in most birds and this slight hypocapnia may itself be the cause of the panting response. If arterial P_{CO_2} is maintained at a constant level during hyperthermia (by inhaling CO_2), then both respiratory frequency and tidal volume increase in the domestic fowl. If the slight hypocapnia of exercise is also prevented, by inhaling CO_2, then the ventilatory response to exercise is similar to that during hyperthermia at normal arterial P_{CO_2} (Brackenbury and Gleeson, 1983). These results make sense, for what they illustrate is that as far as temperature regulation is concerned, an increase in tidal volume is perfectly acceptable, but when the animal breathes air it does cause the excessive removal of CO_2 and therefore alkalosis. This is obviated by the automatic reduction in tidal volume as arterial P_{CO_2} begins to fall. Whether or not the CO_2 receptors in the lungs of birds are important in mediating this response remains to be seen. It is known, however, that bilateral sectioning of the vagus nerves (Figure 3.5) reduces thermal panting in pigeons and abolishes it in fowl, duck and quail.

The respiratory response to hyperthermia is not only affected by breathing gas containing CO_2, it is also affected by low O_2 (hypoxia). When breathing air (P_{O_2} 19.3 kPa), respiratory frequency reaches 260 breaths min^{-1} and tidal volume is 20 ml in domestic ducks at an environmental temperature of 35°C. When breathing a hypoxic gas mixture (P_{O_2} 11 kPa), the corresponding values are 162 breaths min^{-1} and 19 ml, and body temperature is 0.5°C higher than when the ducks breathe air. It appears that some thermoregulation is forfeited in order to conserve energy expenditure when oxygen supply is limited. Despite the lower *overall* ventilation in hypoxic, heat-stressed ducks, compared with those breathing air, *effective* ventilation of the gas exchange surfaces appears to be greater in the hypoxic birds. Arterial P_{CO_2} is only 2.8 kPa in these birds whereas it is 3.7 kPa in heat-stressed ducks breathing air. This suggests that a greater proportion of the tidal volume is involved in gas exchange in the hypoxic ducks, than in those breathing air, in order to keep arterial P_{O_2} at as high a level as possible. Except, perhaps, when flying at high altitudes, it is doubtful if birds would experience heat stress and hypoxia at the same time (remember that body temperature increases during flapping flight even at low environmental temperatures).

4.5.7 *Nervous mechanisms*

The role of the preoptic/anterior hypothalamic (POAH) region of the brain stem, in the response to increased body temperature, seems to vary from species to species. In the Adélie Penguin, warming the POAH is equally as effective as warming the spinal cord in controlling blood flow to the skin and metabolic heat production, a situation similar to that in mammals. Heating the hypothalamus of the Emu causes vasodilatation in the skin and a drop in colonic temperature. In ducks there is a weak response to heating the hypothalamus, causing a reduction in metabolic heat production at cold and thermoneutral ambient temperatures and an activation of panting at warm ambient temperatures. In quail, there is no effect on metabolic rate when POAH is warmed at a range of environmental temperatures, but at the higher temperatures (30°C) there is vasodilatation in the skin when the temperature of the POAH is increased. There is no effect on respiratory frequency in pigeons at a thermoneutral ambient temperature if the POAH is warmed, whereas at high ambient temperatures, warming the POAH actually inhibits panting until body temperature begins to rise.

Bearing these differences between pigeons and ducks in mind, it is interesting to note that, in pigeons, the full panting response (i.e. similar to that during whole body warming) is initiated when the spinal cord alone is warmed, although the threshold for the response is dependent upon temperature in other parts of the body. In ducks, on the other hand, the full panting response is not elicited by warming the spinal cord alone but it may be if the hypothalamus and the spinal cord are warmed together. Heating the spinal cord alone in ducks does, however, induce an increase in carotid blood flow which is close to the full response. There are neurones in the POAH in pigeons whose activity changes in response to warming the spine, so it has been suggested that the POAH integrates signals from the spinal cord, and elsewhere, and initiates thermoregulatory behaviour, such as panting, although it may be insensitive to temperature changes itself (Rautenberg *et al.*, 1978).

The POAH is not the only region of the brain concerned with panting. As long ago as 1936, von Saalfield described a 'panting centre' in the mid-brain of the pigeon. This is now known to be in the dorsal mid-brain and a similar region has been found in the domestic fowl. The importance of this part of the mid-brain is demonstrated by the fact that its destruction eliminates panting altogether (Richards and Avery, 1978). We do not know how the hypothalamus and dorsal mid-brain interact in the control of the respiratory response to heat stress in birds, but a study of the subject is long overdue.

4.6 Birds compared with mammals

The maintenance of deep body temperature several degrees above that of the environment by way of the internal generation of heat (endothermy) is an obvious similarity between birds and mammals, but there are clear differences between the two groups with respect to the level of temperature maintained and to the thermosensitivity of the preoptic/anterior hypothalamic region of the brain. The higher core temperature in birds may be related to the high energy requirements of flight. As already mentioned, core temperature of flightless birds and those which predominantly glide, e.g. albatrosses and petrels, is close to that of mammals (although in bats it is the same as in other mammals).

Caputa (1984), elaborating on a proposal made earlier by Snapp *et al.* (1977), postulates that a high core temperature in birds that perform flapping flight is important for instantaneous take-off and that further elevation of body temperature during flight increases muscular efficiency. However, brain temperature is permanently maintained at a lower value than core temperature in these birds because the brain is more sensitive to high temperatures than the rest of the body. He points out that in mammals and in flightless birds at normal body temperature there is little, if any, difference between brain and core temperature, whereas a difference is apparent if body temperature rises. He argues that as birds evolved and brain temperature was permanently lower than core temperature, it was not appropriate for hypothalamic temperature to act as an indicator of body temperature as a whole; hence the lack of sensitivity to warming of the POAH in birds that fly. In those birds that have lost the power of flight there has been a reduction in body temperature and the hypothalamus has (re)acquired a degree of sensitivity to warming.

As is often the case in Biology, it is possible to find examples which confound such attractive proposals. In this instance, there is the weak sensitivity to warming of the hypothalamus of ducks. Nonetheless, the comparative approach to body temperature and its regulation in birds and mammals could be a fascinating area for future research.

CHAPTER FIVE

OSMOREGULATION

5.1 General considerations

Birds maintain a constant internal environment in terms of salt and water content. Such is a prerequisite for survival and is a feature shared with all other vertebrate animals. Salt and water are gained as a result of feeding and drinking (together with water derived from the metabolism of foodstuffs) and lost via the renal–intestinal complex and by the insensible water loss which occurs principally across the pulmonary surface. Additionally, some birds possess an extra-renal route of salt and water balance in the salt glands. These, when present and activated comprise an important organ system which under certain circumstances assumes greater importance than the kidneys. To remain in osmotic equilibrium the two sides of this 'gain and loss' equation must balance.

The problem faced by birds with respect to salt and water balance turns on the relative availability of both. In nature, the two are inextricably linked since salt and water availability depend entirely on diet. Speaking in hypothetical terms the least onerous situation would arise if these two essential components were freely available from independent sources; thus thirst could be satisfied by free access to a supply of water, and hunger for salt, likewise from a supply of salts. But such is never the case and physiological problems are posed to the animal when an excess of either one or the other characterizes the diet of the animal. Examples of such situations will illustrate these points. In some grain-eating birds there is little water gain from their food and the premium on water is exacerbated in heat stress when evaporative water loss for cooling purposes is also high (see 4.5.4(b)). Under such conditions the conservation of water at the kidney and the integrative segment of the gut (see 5.5.4) is insufficient to maintain water balance. In the absence of access to drinking water such birds will dehydrate and die. By contrast, carnivorous birds feeding on other vertebrates, faced with the same environmental pressures, cope adequately without access to drinking water because the preformed water in food

Table 5.1 Values listed in the table refer to a small *Accipiter* of 150 g body wt, tolerating moderate conditions of temperature and humidity (c. 20°–25°C and 30–40% relative humidity) and it can be seen that the bird can remain in positive water balance by utilizing the salt glands to balance the osmotic equation. The critical factor is the rate of insensible water loss by evaporation. In a falconiform bird the limit of the amount of water that can be lost by evaporation is determined by the concentrating capacity of the salt glands for Na^+ and of the kidney for K^+. The figures in boldface relate to an animal in which the evaporation has been increased to 10%, normal under desert conditions of high temperature and low humidity, then only 10 g of water is available for urine formation and salt-gland secretion. To cope with this environmental stress K^+ concentration in urine would have to rise to 126 mmol l^{-1} and the balance of Na^+ (1.05 mmol) not excreted in the urine would be excreted via the salt glands in 1050 μl of water at a concentration of 1000 mmol l^{-1}. Although these considerations do not result in an exact balance of water and electrolytes internalized and excreted, they do form a feasible strategy for holding the bird in positive water balance, given the putative concentrating capacities of the salt glands and kidneys. These data emphasize the obvious adaptive advantage in terms of water conservation of a functional salt gland for terrestrial carnivorous birds that are not drinking. (Modified after Cade and Greenwald, 1966.)

Item considered	Computation or measurement	Value (based on mouse carcass)
(1) Total body weight	Direct weighing	150 g
(2) Wet weight of daily food	Direct weighing	30 g
(3) Water content of food	66.6%	20 g
(4) Dry weight of food	(2) minus (3)	**10 g**
(5) Oxidative water from 10 g of protein	(4) × 0.449 (based on uric acid end product)	5 g
(6) Total water from food	(3) plus (5)	25 g
(7) Evaporative water loss	2.5% body wt per day	3.75 g
	10% body wt per day	**15.00 g**
(8) Water available for excretion	(6) minus (7)	21.25 g
(9) Total Na^+ content of food	60 mmol kg^{-1} wet wt	1.8 mmol
(10) Average Na^+ content in urine	Flame photometer	75 mmol l^{-1}
(11) Estimated total Na^+ excreted by kidney per day	(8) × (10)	1.59 mmol
(12) Excess Na^+ to be excreted through salt glands	(9) minus (11)	0.21 mmol
(13) Vol. salt gland secretion required at given concentration of 1000 mmol l^{-1}		210 μl
		10.00 g
		0.75 mmol
		1.05 mmol
		1050 μl
(14) Estimated total K^+ content of food	42 mmol kg^{-1} wet wt	1.26 mmol
(15) Average K^+ content in urine	Flame photometry	50 mmol l^{-1} **126 mmol l^{-1}**
(16) Estimated total K^+ excreted by kidney	(15) × (8)	1.06 mmol

(approximately 60–70% by weight) provides an adequate source, given the powers of conservation of water in the kidney and the integrative segment of the gut, supplemented when necessary by functional salt glands (Table 5.1). The final broad category is found in birds which experience a water shortage because of the high concentration of salt in their food and/or drinking water. In its extreme form this situation is illustrated by oceanic birds with no access to fresh water and those living on invertebrate diets which resemble sea water in terms of salt concentrations. If to this is added the problem of water loss due to evaporation at high ambient temperature, then the severity of the aggregate of physiological stresses is quickly realized. Birds in this category rely on an efficient process residing in the salt glands to allow the differential excretion of Na^+ and the retention of water for metabolic purposes. Such extrarenal sites are not confined to marine birds however, since many desert species such as the ostrich and raptors also possess functional salt glands which subserve the same function as those in marine birds and contribute an important complementary role to the kidney and the integrative segment. Herbivorous birds excrete K^+ through their salt glands, since the dietary excess of this ion constitutes the principal osmotic problem.

5.2 Thirst and salt appetite

Fitzsimons (1979) has reviewed this aspect in great detail. Thirst in birds, as in mammals, is an essential physiological requirement to regulate water intake although the sensitivity between bird species varies widely (Kobayashi and Takei, 1982). Osmoreceptors in the hypothalamus and juxtaventricular regions are important monitoring areas in the brain; additionally angiotensin II (the physiologically active component of the renin-angiotensin system) is a potent dipsogenic (drink-inducing) factor and the variability in sensitivity amongst species is attributed to the number and/or affinity of cerebral receptors to angiotensin II; the preoptic area is particularly sensitive in the Japanese Quail. This species can be induced to imbibe 10% of its body weight by the stereotoxic placement of as little as 1 μg of angiotensin into the preoptic area, and such was the immediacy of the drinking response that the crop swelled visibly and water was regurgitated via the mouth, if the birds were handled. Substance P, while inducing a slight dipsogenic effect in the pigeon, inhibits drinking in the Japanese Quail, an effect which is duplicated by enkephalin in this species (Fitzsimons, 1979; Uemura et al., 1983).

Saline preference is evident both from observations in the field where birds on low-salt diet seek to supplement their intake by eating from salt licks provided for cattle, and from experiments offering various concentrations of salt in drinking water; the sensors for salt probably reside in saline-sensitive taste buds.

5.3 Drinking water and evaporative water loss

Bartholomew and Cade (1956 and 1963) in now classical studies contributed in an important way to our understanding of the relationship between water need, relative humidity and body weight. Under standard conditions, they concluded that water consumption matches evaporative water loss, which in turn is a function of body size. Such evidence is presented in Figure 5.1. It can be seen that the relationship between water imbibition (and evaporative loss) and body weight is so significant that it

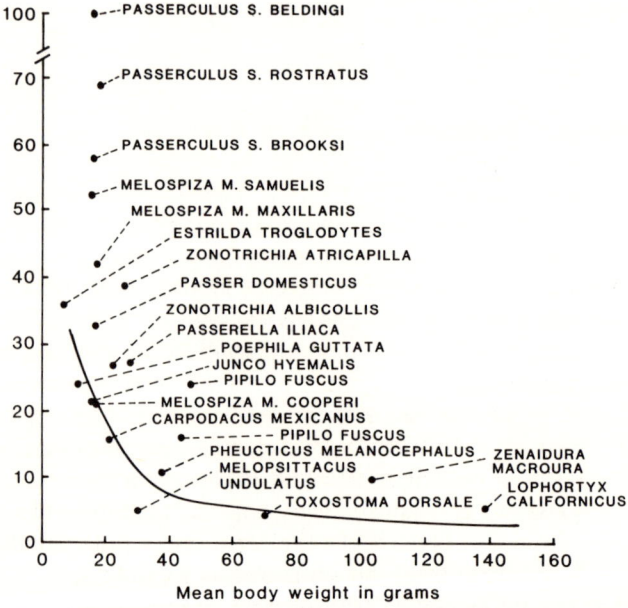

Figure 5.1 Relationship between voluntary water intake and body weight. The relative intake of water at neutral ambient temperature is shown for a number of birds. The solid line indicates the evaporative water loss. Reproduced with permission from Bartholomew and Cade (1963).

masks any other variations between species, especially when desert and non-desert animals are compared.

5.4 Physiological considerations with respect to environmental salinity

The highest concentration of salt in drinking water which can be tolerated without incurring an osmotic penalty is related to the maximum concentrating capacity of the kidney and the integrative segment, after the water loss due to evaporation is subtracted from the total body pool of water. It follows, therefore, that if the evaporative water loss is low the amount of osmotically free water to maintain homeostasis is commensurately greater. Logically, therefore, a small bird with a high metabolic rate, thus producing relatively larger amounts of free water, can tolerate a higher salinity of drinking water. These comments relate to birds without functional salt glands and the presence of this extra-renal mechanism brings an important new dimension to an understanding of their survival value, for the phenomenal concentrating capacity of salt gland 'lifts the pressure' from the kidney and, as will be seen later, has important added survival value both under desert conditions and marine environments.

5.5 Physiological strategies in osmotic survival

5.5.1 *Sites of importance in salt and water balance*

The intestine is the site of primary acquisition of salt and water taken through the mouth. The handling of ingested fluid depends on its tonicity but the end result is the active absorption of Na^+ followed by passive transfer of water. This is shown in Figure 5.2. Whereas most of the reabsorption takes place in the upper parts of the small intestine and in the large intestine, significant amounts of water are reabsorbed in the caeca, ranging from 5–38% of body weight/day. The essential nature of these anatomically distinct organs is in doubt, however, for caecectomy in the fowl does not alter water output by the whole animal. Following exposure to sea water adaptational changes take place in the enhanced ability of the intestine to absorb oral loads; this is shown in Figure 5.3.

The range of environments that a species can tolerate is determined by the physiological properties of the various epithelia, most importantly those found in the kidney and the gut. Certain species can therefore exploit a whole range of environments because of the adaptive plasticity of the

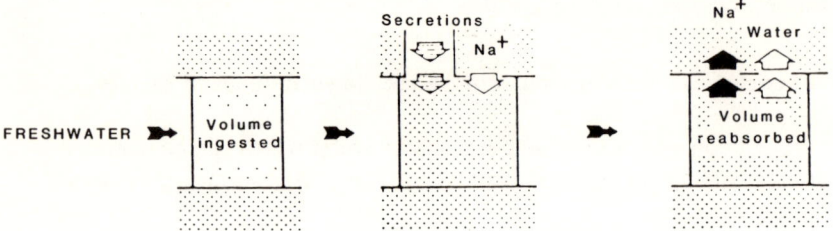

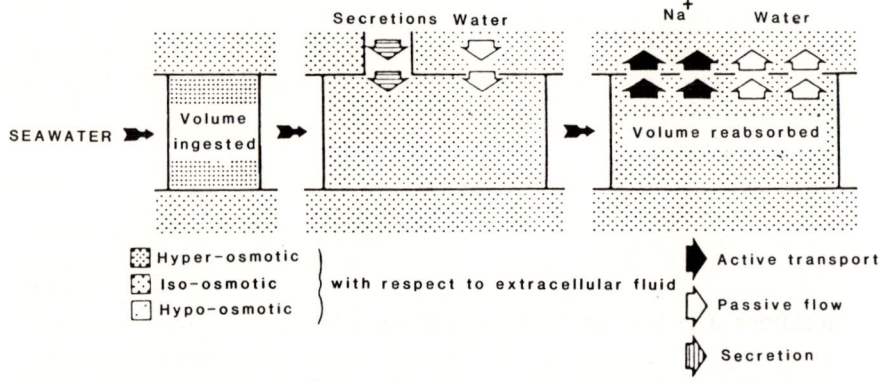

Figure 5.2 A schematic representation of the uptake of water from the gastrointestinal tract of a bird drinking fresh water (upper) and a bird drinking a similar volume of either sea water or hypertonic saline (lower).

In the freshwater situation a hypotonic bolus of water, following arrival in the intestine, has added to it secretions from the gut wall and glandular structures such as the pancreas and liver; the attainment of isotonicity is achieved by passive flow of Na^+. Subsequently, the gut contents are absorbed with passive water flow following the active transport of Na^+.

In birds drinking sea water the initial bolus is hypertonic, secretions are added as in the case of freshwater animals but isotonicity of the gut contents is achieved by passive water flow into the gut. The end result is the same, therefore, for both situations; in the case of the seawater birds the gut content is greater in volume but the pattern of absorption is similar, differing only in the scale of reabsorptive processes. (Taken from Holmes and Phillips, 1985.)

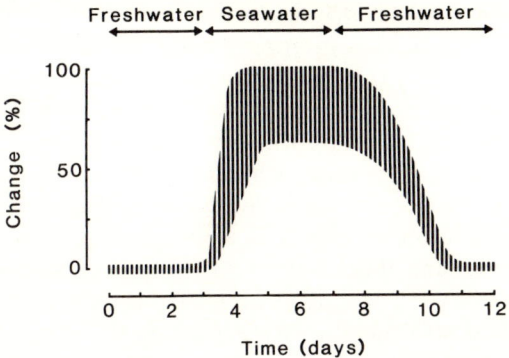

Figure 5.3 The adaptive changes in mucosal water transfer in the small intestine of ducklings during the periods of adaptation to hypertonic saline and fresh drinking-water regimes. The rate of mucosal water transfer is uniform along the length of the small intestine in birds maintained on fresh water but during adaptation to sea water the incremental changes in mucosal transfer are greater in the anterior than in the posterior portions of the small intestine. (Crocker and Holmes, 1971.)

mucosa of the intestine. Such birds can live either in freshwater situations or in conditions of high salt intake—as experienced in marine environments or in species inhabiting the alkaline lake conditions of many inland locations. The adaptive response involves a 'resetting' of the mechanism common to all vertebrates which permits an enhanced uptake of solute-linked water from the gut contents. Because the dominant electrolyte is Na^+ it follows that the rate-limiting characteristic of this part of the gut is the transporting capacity for Na^+. At the same time the volume of water which is absorbed turns on the capacity for Na^+ uptake.

Success in adaptation to marine environments is not, however, related solely to the high rates of solute-linked water transfer across the gut, although this is clearly an advantageous first step. Such is only part of an integrated whole-body response involving the differential excretion of the major electrolytes (Na^+ and Cl^-) via the salt glands, for the uptake of ions through the gut incurs a physiological penalty in that while gaining water for metabolic purposes, the bird gains unacceptably large quantities of salt. The maximum concentrating capacity of the kidneys is such that should the bird rely on the kidney route alone for the excretion of electrolytes, then it follows that on a diet of sea water, more water would be lost than had been ingested and the bird would quickly move into negative water balance and die. There is, however, another aspect of kidney function which is of critical importance in the total economy of water. In birds uric acid is the end product of the metabolism of protein, and as it is a substance of low solubility, a requirement exists, in terms of water, to transport this excretory product down the kidney tubules and ureters to the exterior of the body. While in this context making physiological 'sense', such a requirement would again exacerbate the problem of water loss. This loss is minimized by a mechanism for water conservation which resides in the integrative segment of the gut. The ureteral urine arrives at the urodeal chamber of the cloaca and by mechanical means is refluxed in a retrograde manner first into the coprodeal chamber and then back into the rectal portion of the lower intestine. Here the urine is spaced between the descending faecal mass and the mucosa of the lower intestine. In this position the urine has maximum contact with the reabsorptive surface of the gut. At this site, as in the upper parts of the gut, water reabsorption depends first on Na^+ uptake because the reabsorption of water, unlinked with Na^+, is energetically improbable. Once again, however, the recovery of water carries with it the physiological penalty of 'self-loading with Na^+'. Thus the ingestion of environmental water and the post-tubular handling of urine, in aggregate impose a considerable premium on the body in terms of unwanted Na^+. It is this problem which is met by the activity of the salt glands which through their capacity to produce a highly concentrated fluid, essentially a solution of NaCl, enables the body to excrete sodium and conserve water. It is at this level that final homeostasis is achieved. The evaluation of these interrelated events in gut and kidney has led Schmidt-Nielson to suggest that salt glands have evolved to capitalize on the water-conserving nature of the lower gut.

Salt glands were first described and their physiological role correctly attributed by studies on seawater species. Their occurrence is by no means confined to this ecological group of birds for they have been described in terrestrial species inhabiting hot arid regions where water is at a premium (Cade and Greenwald (1966, Table 5.1) and in species inhabiting environments at the other end of the temperature spectrum where water is at a premium due to its entrapment in snow and ice. Further, salt glands can have great survival value for some species during periods of isolation for example during the incubation of eggs and nestlings of the Bateleur Eagle and the North American Roadrunner which have both been observed to actively secrete from their salt glands.

A dramatic example of a wild species experiencing a natural change in environmental conditions with respect to salt and water intake is described for the Eider duck. The female of the species isolates herself for about one month at the time of egg incubation, during which time she does not eat and rarely drinks. During incubation the salt glands regress and show a distinct reduction in Na^+/K^+-ATPase activity and a reduced capacity to clear a salt load; trends which are reversed when the animal returns to its natural marine environment. During the incubation period the water demands are met by occasional freshwater drinking, and from the metabolism of stored fat and tissue protein, but the stress inherent in these processes leads to high mortality and an unbalanced sex ratio of 1.2 males to each female.

In most birds with salt glands their size and secretory capacity diminishes when the osmotic challenge is removed, but albatrosses may continue to secrete if excited or disturbed, with the result that they may become fatally salt-deficient in captivity without ready access to saline drinking water. There is obvious survival value in the capacity to invoke the full expression of salt-gland activity for those birds facing the environmental stress of shortage of water and excess of salt.

The intrinsic interest in the tissue as a model system for the study of sodium excretion has led many workers to investigate its structure, function and the mechanisms controlling the secretory process.

5.5.2 Discovery of the physiological role of salt glands

Despite the fact that several observations linked salt glands with the osmotic status of birds, it was left to Knut Schmidt-Nielsen and his colleagues in 1957 to provide direct evidence by resorting to the simple device of tasting the salt-gland exudate. Since then there has been an explosion of interest (see reviews by Peaker and Linzell, 1975; Holmes and Phillips, 1985).

5.5.3 Salt glands

(a) *Gross anatomy*. The position of salt glands varies but most commonly they are found in a depression of the skull above or near the eyes. Fossil birds have been found to possess similar depressions so it is a safe presumption that salt glands are ancient structures. Although the glands of many species have been described and variations of detail such as shape, position and drainage have been reported, a single example, that of the Herring Gull, will be described here (Figure 5.4). The glands are paired, flat and crescentic in shape and drained by two ducts suggesting a fusion, during embryological development, of two identical parts. The ducts emerge from the tough connective tissue capsule at the anterior end of the gland, course down through the skull and open separately at the posterior

Figure 5.4 Location of the salt glands in the gull. Note the supra-orbital position of the glands (a). These glands consist of longitudinal lobes about 1 mm in diameter; each lobe has a central canal into which fluid is discharged from the branching secretory tubules arranged radially around it (b). The enlarged diagrammatic sketches (b) and (d) represent a cross-section through the XY plane of the gland (b) and a longitudinal section through the YZ plane (d) of the gland (c). The vascular supply and the microcirculation showing the countercurrent arrangement between blood flow and the flow of secretory fluid along the tubules is also shown. Note the unbranched vessels which penetrate deep into the lobes where, close to the central canal, they break up into smaller vessels which course back in the opposite direction in close association with the secretory tubules. The direction of the blood flow is the reverse of that found in the tubules, forming together the countercurrent system. (Adapted from Schmidt-Nielsen, 1960; Fänge et al., 1958; from Holmes and Phillips, 1985.) (e) Wandering albatross (*Diomedia exulans*) seen on its nesting site on Macquarie Island in the sub-Antarctic. The albatross spends most of its time at sea and with a diet of marine invertebrates is exposed to high salt intake; consequently it has well developed salt glands and in this photograph the droplet of salt gland secretion can be seen on the end of the beak. (Photograph supplied by Mr E. Slater. ARPS, CSIRO Division of Wildlife Research, Canberra.)

end of the *vestibulae concha*. The secretion pools near the external narial openings and is discharged passively to the exterior, dripping off the end of the beak or actively discarded by a vigorous sideways shaking of the head. Again, variations are found and the petrel has a 'water pistol' in the form of a tube above the beak; in this case the expired pulmonary air atomizes the secretion and propels the droplets from this tube. Other species (e.g. the gannets, cormorants and pelicans) have closed external nares and in these species the secretion traverses the roof of the mouth to gain an exit off the tip of the beak (Figure 5.4*e*).

(*b*) *Vascularization*. Blood is delivered to the glands from two branches of the internal carotid artery (internal and external ophthalmic arteries), anastomose ventral to the gland and then emerge to supply blood to the beak. Branches from the bed of anastomotic arteries rise dorsally to penetrate deep into the centre of the gland. Near to the point of the central canal the arterioles reverse direction and cascade radially to form a network of capillaries running in parallel to the secretory tubules. The direction of the blood flow is *opposite* to that of the fluid in the secretory tubule and this simple relationship, by analogy with heat exchange systems (Schmidt-Nielsen, 1960) would be expected to produce an isotonic primary secretion; this point will be returned to later (section 5.3.2 *f*).

(*c*) *Innervation*. The nerve supply to the gland is complex and the details can be sought in Peaker and Linzell (1975). The innervation of the salt gland is predominantly parasympathetic in kind and the release of acetylcholine affects both the vasculature (vasomotor component leading to vasodilatation) and the secretory capacity of the gland (secretomotor component).

(*d*) *Cellular organization*. The architecture of the gland is based on a simple unit of construction, that of a functional secretory tubule which may be single or branched into two or more subsidiary tubules (Figure 5.4). Each tubule ends blindly and it is at this point that the so-called peripheral 'vegetative' cells occur. These cells are small and cuboidal and appear to be actively dividing, especially when the gland is stimulated into activity following a quiescent period. Evidence that the peripheral cells form the stock from which active proliferation of the tubules occur and the ensuing growth of the gland take origin is provided by irradiation studies when the peripheral cells were shown to be the first and most seriously affected by such treatment.

The peripheral cells are already different from the rest of the tubule, which is comprised of principal cells displaying a gradation of increased cellular complexity as the central canal is approached; a characteristic which is magnified following exposure to, for example, hypersaline conditions (Figure 5.5). The appearance of the tubular cells, their position along the duct system and their cytochemical characteristics make them the obvious candidate for the production of the salt-gland secretion. The apices of the radially arranged principal cells form, collectively, the luminal surface of the tubules. Well folded lateral surfaces separate adjacent cells one from the other, forming strikingly complex interstitial channels running from the basal lamina to the apices of the cell where the plasma membranes of

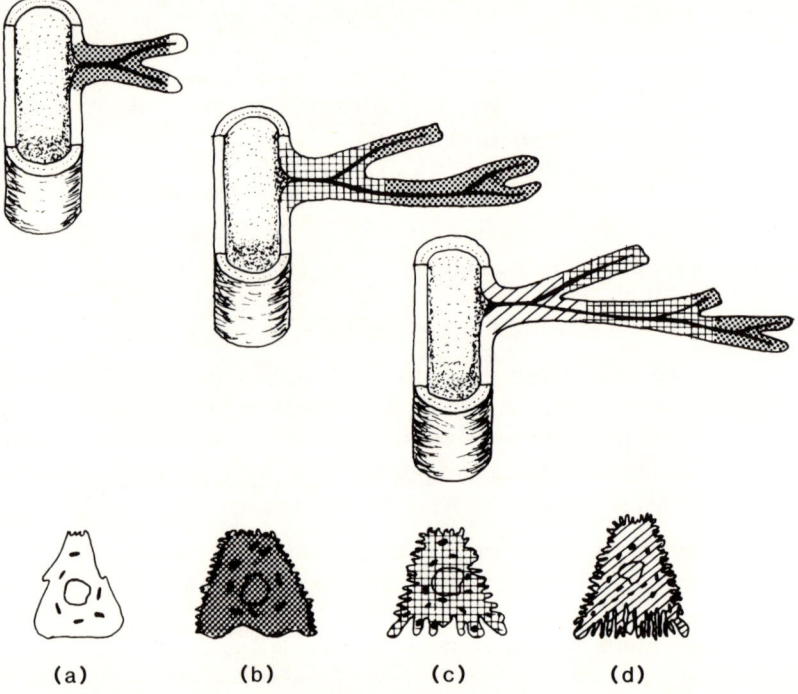

Figure 5.5 A diagram showing the effect of consuming hypertonic saline drinking-water on the development of cellular specialization in the secretory tubule. The peripheral cells (a) located at the blind ends of the tubules show little specialization in their cell surfaces and contain few mitochondria. During the differentiation of the salt gland in response to consuming hypertonic saline, the tubules become elongated and more branched and the principal cells located along the length of the tubule show progressively more complex infoldings of the basal and basolateral membranes. In birds exposed to sea water, the most highly differentiated cells (d) develop at the proximal end of the tubule (cross-hatched region) whereas in birds that consume only fresh water, development stops at the stage of the lesser differentiated cells (type b and c). (Adapted from Ernst and Ellis, 1969; see also Holmes and Phillips, 1985.)

adjacent cells are fused. The junction between cells are permeable to horseradish peroxidase and lanthanum which suggests that movement of fluid between the tubular lumen and the interstitial channels is possible. The complex infolding of the lateral surface of the cells is reflected also by a similar proliferation of the basal plasma membrane resulting in extensive infolding, extending deeply into the cytoplasm (Figure 5.6).

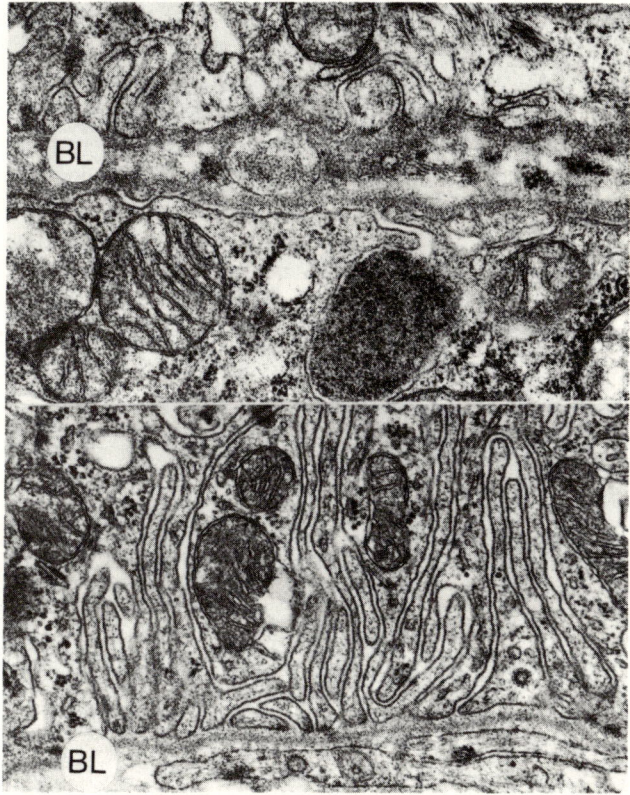

Figure 5.6. Electron micrograph, × 26 000, through the basal portion of a principal salt gland cell from a duck maintained on hypertonic saline drinking water (284 mM NaCl, 6.0 mM KCl). The plasma membrane of the basal surface consists of interdigitating processes that rest on a continuous basal lamina (*BL*). Although the bird was given hypertonic saline drinking water for only 24 h, the interdigitating processes of the plasma membrane have become greatly elaborated. This appearance is in sharp contrast to the inactive cells depicted in the freshwater animals (top photograph). (Cronshaw and Holmes, unpublished, in Holmes, 1972.)

(e) *Differentiation of the stimulated gland.* The remarkable feature of the avian salt gland is that the quiescent gland can be activated within minutes following a rise in blood osmolarity occasioned by an increase in dietary Na^+ intake; thus despite inactivity for long periods, the tissue retains all the functional secretory mechanisms necessary for a rapid response. Persistent exposure leads to a growth in the size of the gland (hypertrophy), of about 60%, accompanied by a progressive differentiation of the principal cells by elaboration of cellular components; this together with a marked increase in cell number (hyperplasia) of about 40% produces a more efficient excretory capability, an overall process which, in the domestic duck, takes about 1–2 weeks. Since the % tissue water remains constant the increase in the salt-gland weight can be attributed to the laying down of additional protein in the form of connective tissue, blood vessels, cellular components and enzyme systems. The increase in protein synthesis is preceded by a striking increase in RNA synthesis, which within one day of exposure can rise threefold and be maintained there so long as the osmotic stress persists (Figure 5.7). The functional development also involves increases in phospholipid content as the baso-lateral cell membranes proliferate, increases in soluble enzymes of the cytoplasm, an increase in mitochondrial count and mitochondrial cytochrome oxidase and succinic dehydrogenase levels. All

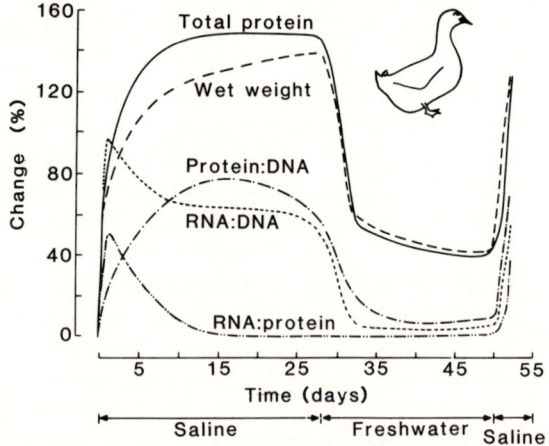

Figure 5.7 Modification in the wet weight and protein content together with the accompanying relative changes in the protein and nucleic acid composition of salt glands from the duck during periods of adaptation to hypertonic saline alternating with fresh drinking-water. These changes are possibly due to the DNA/RNA mediated protein synthesis under the influence of corticosterone. (Holmes and Stewart, 1968; Homes and Phillips, 1985.)

of these parameters are reversed when the saline stress is removed so that the tissue reverts to its presecretory appearance and set of characteristics.

(*f*) *The secretory mechanism.* The earliest work on salt glands included experiments designed to elucidate the mechanism by which the secretion is formed. The first clues came from the use of ouabain (which blocks Na^+-K^+-dependent ATPase) injected in a retrograde fashion into the duct lumen, resulting in an abolition of the positive electrical charge of the luminal contents relative to the tissue and the abolition of the secretion. These observations provide support to the idea that the concentrating mechanism involves a Na^+ pump on the luminal surface of the cells mediated through the action of a Na^+-K^+-dependent ATPase enzyme. Moreover, it was later shown that whereas ATPase activity in quiescent salt glands is sufficient to meet the excretory needs of birds exposed briefly to hypertonic conditions, survival following long-term exposure depends on greatly enhanced levels of activity. The proliferation of the gland, therefore, provides an enhancement of the driving force for the secretion in terms of functional enzymes as well as the structural framework to carry this expanded capability.

Various theories have been advanced to account for the phenomenal concentrating capacity of salt glands and these have recently been reviewed in detail (Holmes and Phillips, 1985). Until recently, three main possibilities existed, viz., active transport of Na^+ across the apical membrane of the principal cells of the tubule and transcellular flow, active transport of Na^+ into the intercellular spaces of the principal cells and the resultant paracellular flow of a hyperosmotic fluid into the lumen, and finally isosmotic fluid secretion by the peripheral cells at the blind ends of the tubules and the reabsorption of water by the principal cells to produce a hyperosmotic secretion (reviewed by Holmes and Phillips, 1985).

The underlying assumption of all these theories is that the highly concentrated exudate which appears at the external nares is elaborated by the tubules of the gland and subsequently remains unmodified. There is no direct evidence for this. Recently, Marshall *et al.* (1985) using X-ray microanalytical methods have been able to measure elemental concentrations of ions at sub-cellular and luminal locations. The results are presented in Figure 5.8 from which it can be seen that post-tubular modification of the primary secretion is of the greatest significance. The results suggest that reabsorption in the main duct system allows concentrations of the final exudate to be controlled by water reabsorption from a primary isosmotic secretion produced at a variable rate by the secretory tubules. The

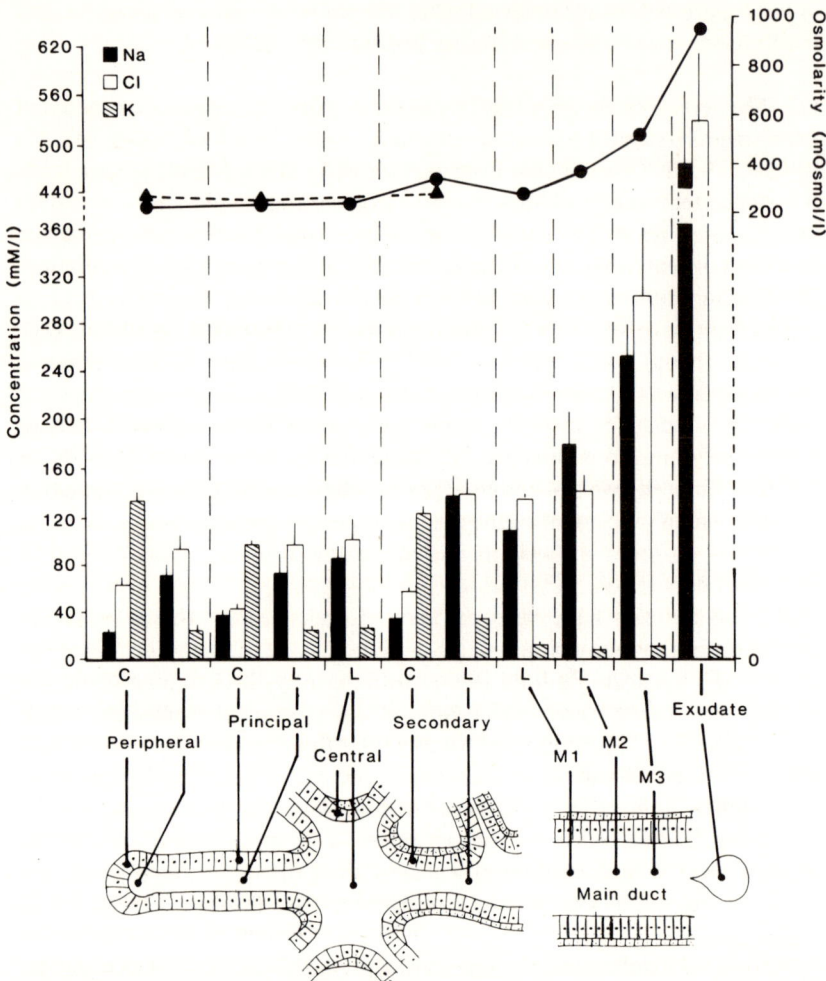

Figure 5.8 A summary of the ion concentrations found at a series of different loci along the duct system of the salt gland in ducks previously adapted to hypertonic saline drinking-water and loaded intravenously with hypertonic saline. Samples of intracellular and luminal fluid were taken at two sites along the secretory tubules, from the central canal into which fluid from the secretory tubules drains directly and from the secondary and main collecting ducts which receive fluid from the central canals. The intracellular ion concentrations of peripheral and principal cells in the secretory tubule and the cells lining the secondary ducts are also shown. These data indicate that the phenomenal concentrating capacity of the salt gland resides in the main duct. (Marshall *et al.*, 1985.)

countercurrent arrangement in the salt gland in fact is ideal for the elaboration of a secretion which is isotonic with plasma.

(*g*) *Initiation and control of secretion.* Birds which possess salt glands but have adequate access to fresh drinking water will exclusively use the integrative segment/renal route of salt and water balance to achieve homeostasis. When challenged with conditions where water is at a premium and the capacity of the integrative segment/renal route to osmoregulate is exceeded, the salt glands will come into play but will cease to secrete when the stress of water shortage is removed.

The rapidity with which the mechanisms responsible for the initiation of the secretion are activated is borne out by the appearance of the exudate at the external nares within minutes of the ingestion of hypertonic drinking water and suggests, because of the short time scale, a neural control mechanism. Such is confirmed by denervation experiments where one or both secretory nerves are cut, leading to unilateral or complete cessation of salt-gland function. The important factor responsible for secretion is the release of acetylcholine from nerve endings which have been observed to terminate near to the secretory tubules and the blood vessels which richly supply the glandular tissue. As a result of acetylcholine release secretion commences but also its vasodilatory action opens up the vascular bed of the gland. As a consequence, enhanced delivery of substrate to supply the energy-intensive metabolic demand of the tissue occurs and the delivery of certain factors such as hormones is satisfied as well as the delivery of electrolytes for excretion.

The determination of the 'strength of the signal' to the gland in terms of acetylcholine release is a result of a complicated interplay of several physiological factors, the relative importance of which depends on the status of the animal in terms of salt and water balance (Simon, 1982). Osmoreceptors have been described both in the region of the heart and the hypothalamus. The input–output coupling in osmoregulation is achieved to a considerable extent by hypothalamic neuronal networks. The osmosensitive afferents from the wall of the 3rd ventricle as well as ascending neurones from the vagal sensory nuclei (in turn linked to vagal osmotic and volume-sensitive afferents), together with inputs from various brain stem regions, have been shown to converge in the paraventricular nucleus. This conclusion has importance, for it is at this site that the release of arginine vasotocin (AVT), the anti-diuretic hormone of birds, occurs. Neurones from the hypothalamic region descend to the parasympathetic brain stem nuclei which then innervate the salt glands via the secretory nerve ganglion (*ganglion ethmoidale*) and the secretory nerve, the branches of which intimately terminate at the level of the secretory tissue. In this connection it is interesting to note that hypothalamic cooling not only blocks AVT release but also inhibits salt-gland activity in penguins and ducks.

Plasma tonicity and its measurement by osmoreceptors is only part of the picture, for volume perception has recently been shown to be of particular importance in the control of osmoregulatory effectors in the duck. The relationship of AVT release to the osmotic 'monitoring' of plasma osmolality by osmoreceptors is reset at a lower level by volume loading both in fresh- and saline-water-adapted animals (Simon-Opperman *et al.*, 1980). The available evidence suggests that extra-vascular volume changes are generally more important than intravascular changes in avian osmoregulation (Simon, 1982). The possibility of stretch receptors recording volume changes in vascular compartments in birds should not be discounted, although no clear evidence for this exists at this time, as it does for mammals.

(*h*) *The role of hormones in salt-gland function.* The initiation of salt-gland secretion, as we have already seen, occurs in response to a neural stimulus, namely the release of acetylcholine, but the maintenance and full expression of the secretory process requires a significant endocrine input. Hormones affect salt-gland function in one of two possible ways; first by direct effects on the gland and secondly by contributing in a variety of ways to the integrity of the internal environment, which is essential if normal homeostasis is to be sustained.

Descriptions of direct effects on salt-gland function by a variety of hormones have appeared in the literature (see reviews by Harvey and Phillips, 1982*b*; Holmes and Phillips, 1985). Much of the evidence can be interpreted as effects which are secondary in nature, but there is good evidence for a primary effect on salt-gland function for both AVT and corticosterone. All other hormones seem to subserve a secondary or supportive role, but are no less important in the overall maintenance of homeostasis because of that.

We have already noted a relationship between the osmotic status of the bird and the level of secretion and release of AVT (reviewed by Simon, 1982). AVT and its near relative oxytocin are the only hormones which can induce secretion from the salt gland in the absence of a salt load. Oxytocin produces a secretion of normal composition while AVT produces a secretion with depressed Na^+ concentrations. A direct effect for AVT is suggested by experiments showing that smaller doses are required to induce secretion when introduced into the carotid artery near to the gland than when injected into the general circulation. As the secretion elicited in this way is diluted, the possibility exists that AVT affects water permeability in the gland. An alternate role for AVT has been described through its possible effects on the kidney which, by altering the internal medium of the body, secondarily affect salt-gland function.

Hormones from the adrenal gland were the first to be examined with respect to a possible role in salt-gland function. In support of their involvement can be listed the following evidence.

(i) The size of the adrenal gland increases with the degree of exposure to hypersaline conditions and the increase occurs in the central zone which is known to be the site of corticosterone synthesis and secretion.
(ii) Corticosterone synthesis increases and circulating levels of corticosterone are elevated following salt loading.
(iii) Corticosterone is rapidly metabolized to 11-dehydrocorticosterone by salt-gland tissue.
(iv) Hypophysectomy reduces, and adrenalectomy abolishes, the ability of salt loaded birds to respond normally with respect to salt-gland secretion, and replacement therapy with ACTH, cortisol or corticosterone restores this capacity.
(v) Corticosterone is preferentially taken up by salt-gland tissue on to intracellular sites.

The salt gland is likely to be a target organ for corticosterone action, since two types of specific corticosteroid-binding macromolecules have been isolated from the cytosolic fraction of activated salt-gland cells; one type of receptor has a high affinity but a low capacity to bind hormone, while the corresponding characteristics of the other type of receptor reflect a low affinity and a high capacity. Each type of receptor has a greater affinity for corticosterone than for hormones with predominantly mineralocorticoid properties, such as aldosterone and deoxycorticosterone.

The cytosolic and nuclear fractions isolated from the adjacent Harderian gland, and from more remote tissues such as lung and liver, do not possess the abilities either to bind or to metabolize significant amounts of labelled corticosterone. The specific properties of the receptor complexes isolated from the salt gland, therefore, strongly suggest that corticosterone may be involved in transcriptional events leading to the full functioning of the salt gland in birds consuming hyperosmotic drinking water. The exact nature of the mechanisms whereby either corticosterone or 11-dehydrocorticos-

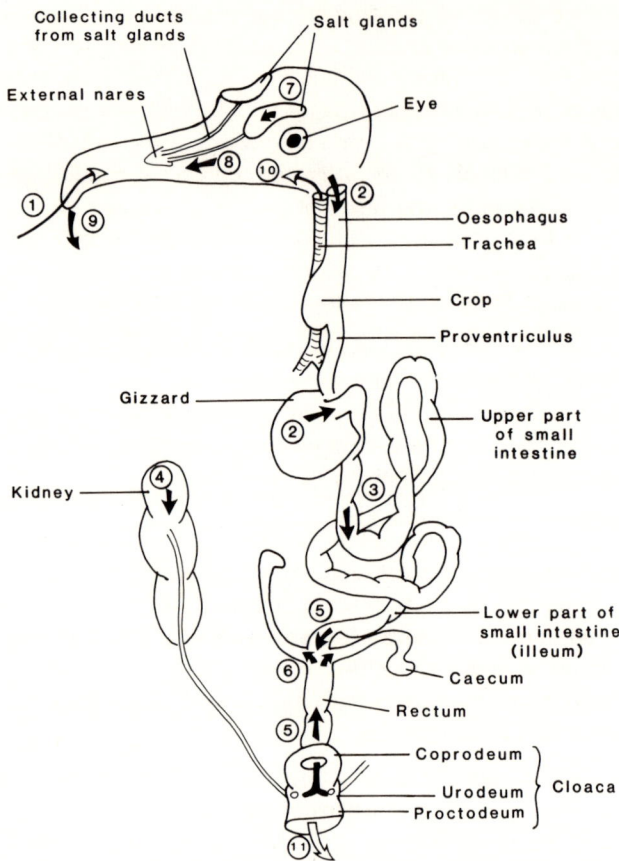

Figure 5.9 Overall schema indicating location of main physiological events in achieving homeostasis in terms of salt and water balance in birds (not drawn to scale.)

Site	Major physiological event	Possible controlling factors
1. Mouth: food and water ingested	Taste Modification of food and water intake (appetite)	Prolactin Angiotensin II
2. Food and water reach upper intestine	Secretions added Beginning of process to achieve isotonicity	
3. Small intestine	Equalization completed (see Figure 5.2)	Thyroxine/T3

	Gut uptake	Corticosterone Aldosterone Prolactin
4. Kidney	GFR adjusted	Arginine vasotocin (AVT) Corticosterone
	Antidiuresis Isosmotic/ Hyperosmotic urine	AVT and prolactin AVT aldosterone corticosterone Mesotocin?
5. Lower segment small intestine + ureteral contents mix in rectum	Reabsorption of Na$^+$ followed by water from isosmotic chyme following equalization process	Prolactin Corticosterone Aldosterone
6. Caeca	Amplification of events in (5).	?
7. Salt gland	Copious production of isosmotic fluid	Acetylcholine Corticosterone AVT
8. Collecting ducts	Efficient concentrating mechanism to produce hypertonic exudate	?
9. Beak	Salt gland exudate drips off beak	—
10. Lungs and respiratory tract	Insensible water loss from lungs	—
11. Cloaca	Complementary events to (5).	

terone (to which it is rapidly converted) may evoke these changes, however, is not yet known. Nevertheless, the general characteristics of this steroid recognition system are quite similar to those found in other known glucocorticoid-regulated mechanisms.

Taken together the evidence suggests a direct action of corticosterone in the process of the adaptational responses of the salt gland to sustained demand for activity. Thus corticosterone may be the candidate for the steroid-induced, RNA-mediated, protein synthesis required for the production of enhanced levels of metabolic enzymes including Na^+–K^+-dependent ATPase, and may be responsible for the growth through hypertrophy and hyperplasia of the tubular elements, connective tissue framework and attendant blood supply to the gland.

Such a thesis is supported by recent unpublished studies of Phillips, Marshall and Wright who carried out a simple field experiment on Noddy Terns on Heron Island in the Great Barrier Reef. These terns have no access to fresh water and are exposed to long hours of thermal gain as they incubate their eggs in tropical conditions; salt-gland activity is evident from the exudate that can be easily observed. In a controlled experiment the injection of corticosterone resulted in a significant increase in salt gland size. Corticosterone may therefore be the elusive growth factor in the proliferation of salt-gland tissue.

While the hormones of the posterior lobe of the pituitary and adrenocortical hormones have alone been assigned a direct role on salt-gland function, a whole variety of other hormones are secondarily involved. Appetite for food, salt and water is influenced by prolactin and angiotensin II. Gut uptake depends on corticosterone and thyroxine, kidney function on AVT, corticosterone and aldosterone, post-renal modification of urine in the lower intestine on prolactin, aldosterone and possibly corticosterone. So this gamut of hormonal events combine to ensure the appropriate physiological response in maintaining the constancy of the internal environment of the body (Figure 5.9).

5.5.4 The lower intestine as an integrator of renal and intestinal excretion

Birds have the capacity for post-renal alteration of the composition of urine arriving via the ureters at the urodeum of the cloaca; this is achieved by the merging of the intestinal and ureteral fluids in the integrative segment of the intestine. The integrative segment is defined as that part of the intestine in which intestinal fluid and urine mix (achieved by a backwards refluxing of the cloacal contents into the intestine) in the presence of a transporting epithelia capable of altering the composition of the luminal fluid; this modification may be regulated to satisfy the metabolic demands of the body. The extent of this integrative segment is determined by the furthest point urine penetrates in a retrograde fashion up the intestine and embraces not only the coprodeum of the cloaca and the rectum (sometimes referred to as the colon) but also the intestinal caeca (where they occur) and even the lower (posterior) parts of the ileum (Figure 5.9). It is important to realize that the events which take place in the integrative segment are determined not only by what is delivered to this segment, which is of renal origin, but also the volume and composition of the fluid of intestinal origin at the ileum/rectal junction where the two volumes of fluid meet and mix. This aspect of salt and water balance has been expertly reviewed by Thomas (1982, 1983) who goes on to conclude that 'despite the substantial turnover of water and Na^+ the intestine anterior to the integrative segment performs little net absorption or secretion of these substances and that it is therefore not important in regulating salt and water balance, or least as far as fowls are concerned. The corollary is that the ileum delivers water and Na^+ to the integrative segment in amounts and concentrations similar to those consumed. On the other hand, the small intestine performs a net absorption of potassium and calcium and may have an important regulatory function with respect to calcium balance'. As stated, this conclusion is based largely on studies on the fowl and need not necessarily hold for birds with salt glands and drinking sea water for, as we have recognized earlier, hypertonic drinking water is appropriately modified in order to achieve isotonicity in advance of absorption of the gut contents.

The relative proportions of the constituent parts of the integrative segment vary enormously; the rectum as a percentage of the total intestinal length can range from as high as 68% in the ostrich to as low as 1% in the thrush, reflecting presumably the degree of severity of the environment experienced by these species. An interesting observation in this connection has been made by Amanova (1975) who compared desert-living sparrows and two related species which share the same environment but are wholly dependent on water supplies acquired at human settlements. The desert-adapted species have a marked development of the rectal epithelia and a capacity for enhanced fluid uptake from the lumen of the rectum compared with the species with free access to water—a good example of a natural controlled experiment in the wild.

(a) *Role of the integrative segment.* Greatest attention has been focused on the coprodeum and rectum, but the caeca are remarkable in that they play an important role in water conservation. Caeca do not, however, occur in all species and, further, birds that possess them can, following caecal occlusion (or caecectomy) transfer, without apparent ill effect, this important role to other parts of the lower intestine in a compensatory fashion, provided there is full access to salt and water. The achievement of homeostasis following these procedures demonstrates the compensatory reserve capacity of other parts of the intestine and explains in large measure that the presence or absence of caeca is not obviously related to either diet or habitat. Whether this would hold equally true for birds subjected to dehydration has not been tested. The caeca are the site of considerable production of volatile fatty acids (VFA) produced as a result of microbial fermentation, and their water-conserving role is linked both to Na^+ and VFA absorption at this site.

(b) *Cloacal and rectal function.* Two approaches to the establishment of the relative contribution of the cloaca and rectal portions have been pursued, by isolating parts of the intestine in experiments *in vitro* and relating these findings to a companion set of data derived from whole or *in-vitro* experiments by perfusion of parts of the gut. Thus perfusion studies of the rectal and cloacal segments of the intestine of the fowl and the Galah (the seed-eating parrot from the arid and semi-arid parts of Australia) taken together with studies of *in-vitro* strips of rectal and coprodeal epithelium have indicated that the coprodeum alone, of the chambers of the cloaca, is functional in transmural exchanges which contribute to salt and water balance. Irrigation of the segment as a whole with fluid which resembles ureteral fluid in composition results in net absorption of Na^+, NH_4^+ Cl^- and phosphate and outwards secretion of K^+. Osmotic fluxes across this segment of the gut occur but important also is the solute-(Na^+)-linked water absorption. For example, net water absorption can occur with luminal osmolarities very much higher than plasma, and it would seem that such net transport takes place against large lumina–plasma differences in birds such as the Galah which experience more severe environmental conditions of lack of water than would be the case, for example, in the fowl. This conclusion is reinforced by experiments on the fowl where solute-linked water flow is enhanced under conditions of enforced dehydration.

The mechanism by which the integrative segment responds to dehydration is unknown; arginine vasotocin at levels sufficient to cause moderate antidiuresis did not change Na^+ or K^+ transport in the recto-coprodeal segment of the fowl.

Manipulation of Na^+ levels to cause depletion increases the capacity of the fowl's integrative segment to absorb Na^+ and hence water. This may be mediated by aldosterone, as would be expected *a priori*, but since aldosterone injections only partially reproduce the many effects in the lower intestine observed after Na^+ depletion, another factor, possibly prolactin, may act in concert with aldosterone (Figure 5.9).

Studies *in vitro* have contributed further to our understanding of the relative importance of the rectum and coprodeum and our current understanding allows for a tentative generalized view, viz., that the recto-coprodeal function in the fowl is characterized by active Na^+ absorption correlated with K^+ secretion mediated by a ouabain-sensitive $Na^+ - K^+$ exchange system at the basolateral cellular sites. The rectum, but not the coprodeum, is important for solute-linked water absorption at rates which vary depending on the state of hydration. Further, the caeca are quantitatively more important than the rectum but whether they are as sensitive to hydration status is not known.

The essential physiological roles played by the integrative segment are elegantly reviewed by Skadhauge (1981) and Thomas (1982, 1983). Having examined all the evidence Thomas (1982) concludes

> All the available information points towards a quantitatively important role of the caeca (where present), the rectum and the coprodeum in modifying the composition of intestinal fluid and ureteral urine before excretion, and thus supports the concept of the integrative segment. Moreover, those portions of the integrative segment which have been examined in respect of their regulation (i.e. the rectum and coprodeum) clearly show that their capacity for net transfer of salts and water is modified appropriately in response to the animal's prevailing metabolic requirements. Thus the avian integrative segment has the qualitative attributes and quantitative importance to be expected of a major homeostatic regulator of salt and water metabolism. This is not surprising in view of the segment's strategic position at the nexus of renal and intestinal flow and just prior to the common point of excretion. The reason it has taken so long to recognise the functional importance of the integrative segment for what it is seems to have more to do with the historical preoccupation of physiologists with mammals than with any subtlety, complexity or marginal value of the system itself.

Although the various parts of the integrative segment act in concert and the relative contribution of its parts depend on the volume and composition of the admixture of fluid delivered both from that ingested by the mouth (suitably modified by the upper intestine) and from the retrograde flow of ureteral fluid, the overall metabolic economy of the bird is best served by demands which minimize adjustments to the luminal fluid in bringing it to isotonicity. Thus the production of a hypertonic urine would be counter-productive in terms of water economy where the integrative segment is concerned for this part of the array of homeostatic mechanisms works most efficiently under conditions of isotonicity. This important point will be returned to later (section 5.5.5).

5.5.5 *The kidney*

(*a*) *Anatomy and morphology.* The avian kidney possesses interesting structural features which bear resemblances to the mammalian and lower vertebrate kidneys. In gross anatomical terms the kidney resembles that of the chelonian and saurian reptiles. The blood supply is not segmented, as in many lower vertebrates, but it is supplied with more than one artery, as in mammals. The avian kidney is endowed with a second supply of blood from a functional portal system and in this resembles once again the lower vertebrates. Birds, however, stand alone in the possession of a smooth muscle valve which controls the amount of venous blood which traverses the kidney tissue; the venous drainage is by a common renal vein. In its internal structure, the avian kidney exhibits characteristics of the reptilian and amphibian kidney combined with some of the features of the mammalian kidney. The cortical area resembles the kidney of lower vertebrates and the medullary area, that of mammals. The cortical nephrons are reptilian-type (RT) nephrons, simple in arrangement, being tubules folded upon themselves 3 or 4 times without loops of Henle and emptying at right angles into collecting ducts. The RT nephrons radiate about the axis of the central efferent vein and form cylindrical cortical lobules. Where the cortical lobules meet are larger, more complex nephrons which do possess loops of Henle, and these resemble the mammalian type of nephron (MT nephrons). The loops of Henle of the MT nephrons, the *vasa recta* which take their origins from the efferent asterioles of the glomeruli

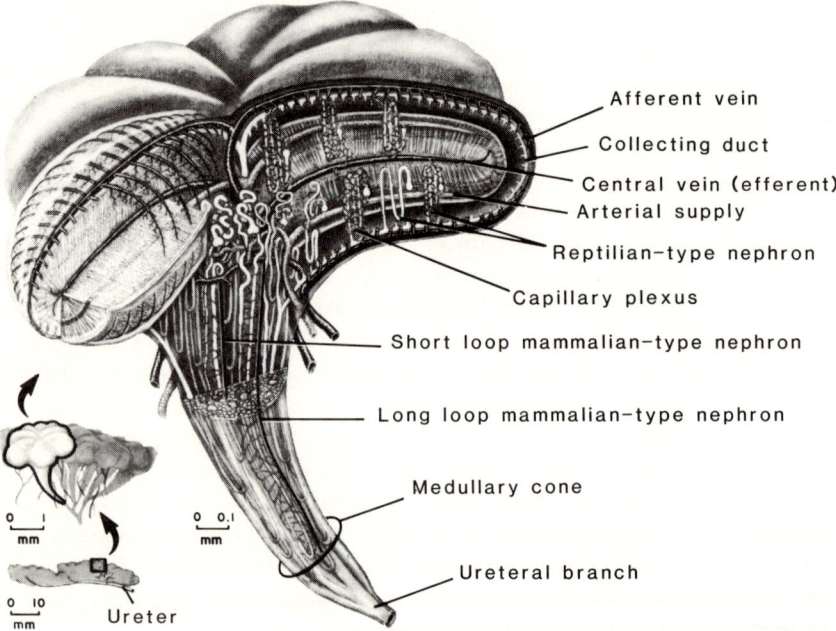

Figure 5.10 Illustration of a section of the avian kidney showing types of nephrons present, their relative positions in the kidney and their relationship to other intrarenal structures. The whole kidney is illustrated in the lower left of the figure. Two enlargements are shown, each showing greater detail than the previous one. (From Braun and Dantzler, 1972.)

together with the collecting ducts draining the RT and MT nephrons, are bound together by connective tissue to form the medullary cones (Figure 5.10). The detailed circulation of the kidney is reviewed by Braun (1982) and Skadhauge (1981).

The water-conserving capacity of the avian kidney takes place in the medullary cones by the simultaneous excretion of excess ions. This is achieved by a countercurrent multiplier system which employs Na^+ almost exclusively, whereas in mammals urea, as well as Na^+, are important components which provide the osmotic gradient in roughly equal amounts. The amount of medullary tissue, not surprisingly, dictates the degree by which kidneys can concentrate above the osmolarity of plasma. In general the concentrating ability of the bird kidney falls short of that of the mammalian kidney, but in certain desert species, for example the Zebra Finch and the Galah, a 2.5–3-fold difference in urine over plasma osmolarity has been observed (Skadhauge, 1974) and the Savannah Sparrow when drinking saline reaches a 6-fold difference. It must be emphasized, however, that the involvement of the integrative segment of the lower intestine ideally requires an isosmotic ureteral effluent. This would minimize the metabolic demands of producing a hypertonic urine, only to have this reversed by the equalization process (between intestinal fluid and plasma) of the coprodeum and rectum to achieve isotonicity in advance of absorption of Na^+ and water (see section 5.5.4).

Interesting variations have been described in the way certain birds handle osmotic stress. For example, the Desert Quail, found in semi-arid locations, does not form a hypertonic urine following dehydration or salt loading but accommodates the osmotic stress by increases in plasma and urine osmolarity. This strategy therefore produces osmolarities in the extracellular compartment which are quite outside the normal physiological range (Braun and Dantzler, 1972) and must lead to consequential changes in the intracellular environment. In contrast, the Starling, when similarly challenged, adapts by maintaining plasma osmolarity by the device of utilization of body water to excrete, via the kidneys, the excess solute load in an increased volume of urine but with an unchanged glomerular filtration rate (GFR) (Braun, 1978). It has been suggested (Braun, 1982) that the differences observed in these two species reflect their respective capacities for flight; the quail is not capable of extended journeys in search of surface water, whereas the Starling is. As Braun comments (Braun, 1982) these comparative studies suggest that examination of urine to plasma ratios in birds have less relevance than in mammals because (a) plasma osmolarity is very much more labile in birds, tending to rise during dehydration and fall with hydration; (b) the overall metabolic economy of the body requires an isotonic rather than a hypertonic urine because of the integrative segment of the gut acting as an 'extension' of kidney function; and (c) uric acid is less osmotically active than urea because of its low solubility.

(b) Glomerular filtration rate (GFR). Described in general terms, birds stand between lower vertebrates and mammals in the variability of GFR. Mammals are remarkably constant, whereas birds are less so, with the greatest variability found in birds in which free access to water poses a continual problem. Birds with functional salt glands have elevated GFRs to clear the expanded extracellular fluid volume as a result of drinking sea water, with the stepwise handling of salt and water balance by the subsequent events in the integrative segment and the salt glands.

(c) Nephron intermittency. Mention has been made of the two types of RT and MT nephrons. Osmotic stress in gallinaceous birds leads to a decrease in the number of filtering nephrons, a process called nephron intermittency, a phenomenon principally confined to the RT nephrons. When an extreme salt load is administered to the Desert Quail, the RT nephrons appear to stop filtering, an effect which can be duplicated by the intravenous administration of AVT. This hormone acts on GFR at the level of the afferent arteriole. AVT therefore has a dual role on the kidney, that of controlling nephron intermittency on the one hand, and the more traditional role of regulating the permeability of the collecting ducts to water on the other.

The other important endocrine factors in bird kidney function include the control of aldosterone by angiotensin; aldosterone production is also sensitive to ACTH in ducks but not in fowls. Also, as in mammals, the sodium status of the bird determines aldosterone production rates. Candidates for target sites for aldosterone action are the rectum and the kidney tubule with a paramount role for the rectum. In this connection, colostomy in the duck (Wright *et al.*, 1982), where the kidney and rectal effluents can be collected separately, results in no changes as far as the ureteral fluid is concerned between control and colostomized birds. The important observation that aldosterone levels are markedly raised following colostomy would, therefore, suggest a role for aldosterone in stimulating active sodium transport in the rectum to compensate for the lack of availability of the ureteral fluid to the lower intestine sites due to the surgical procedures.

There is, however, an alternative method of collecting ureteral and integrative segment luminal fluid which does not involve colostomy (Thomas et al., 1984). These studies indicate that penetration of ureteral fluid into the lower intestine is not an obligatory step for birds such as the Chukar and the Sand Partridge, an observation previously reported for the ostrich (Skadhauge, 1981). Thomas et al. emphasized what has already been mentioned earlier, that if ureteral urine is sufficiently hypertonic then refluxing into the integrative segment could result in increased net water loss if the osmotic dilution of the gut contents in the 'equalization' process exceeds the solute-linked water uptake in this segment. Obviously, this will depend on the span of time the ureteral fluid rests in the gut before expulsion to the exterior. If it is short, then osmotic loss of water into the lumen will predominate over the reabsorptive solute linked uptake. Viewed in another way, as Thomas et al. state, 'when birds are fully hydrated or when the ureteral urine is concentrated maximally, there may be little benefit for their water economy to be had from urinary reflux up the intestine'. Interesting, too, are the observations (Skadhauge, 1981) that the threshold of sensitivity to aldosterone is greater in rectal than cloacal sites—why this should be so is not clear. What is clear is that there is a reciprocal relationship between aldosterone and prolactin in birds adapting to high Na^+ diets, and prolactin may actually antagonize aldosterone during such changes in Na^+ status. Such would make good physiological sense in animals without salt glands by contributing to the prevention of Na^+ accumulation in the body. Conversely, in birds with salt glands which can differentially excrete Na^+, and when aldosterone levels are low because of salt loading, prolactin may have a role in promoting solute-linked water uptake in the integrative segment (Ensor, 1978). The stimulus for prolactin release may be, as for the release of AVT, elevated Na^+ concentrations and blood osmolarity, for under such conditions the plasma hormone concentrations of the two hormones rise in parallel fashion.

(d) *Nitrogenous excretion.* The major excretory product of nitrogen metabolism is uric acid synthesized in the liver and/or kidneys. Most derives from secretion by the tubules but some, about 10%, is filtered. The concentrations of uric acid and urates considerably exceed their low solubilities and as a consequence they exist in colloidal forms (80%) which are stabilized by mucoid secretions. Consequently, they contribute little to the osmolality of the formed urine and the osmotic force may be further reduced by the co-precipitation of part of the dissolved cationic pool—up to 75% for Na^+, 94% for K^+ and 32% of Ca^{2+} and Mg^{2+} depending of course on the proportion of urate precipitated (Skadhauge, 1981; Thomas,

1982). Further, as Braun (1982) has stated, if uric acid is excreted as the monobasic salt the osmotic activity of the urine would be reduced by an amount equivalent to the ions associated with uric acid. What now seems to occur is an amplification of this binding process in excess of that which is accountable for by equimolar ratios of ions and uric acid. Braun (1982) cites evidence that uric acid can be further modified by bacterial action in the cloaca to free the sodium for reabsorption and in this way contributes to water economy.

5.6 Behavioural aspects of osmoregulation

Many species have evolved feeding strategies which enable them to remain inactive during the hottest periods of the day; this minimizes thermal gain and minimizes the need to dissipate metabolic heat. Journeys to areas of surface water are, in general terms, restricted to the cooler early morning and late evening. Birds such as the Sand Grouse tend to fly longer distances when disturbed in the cooler parts of the day than in the hotter parts (Thomas and Robin, 1977). Birds in hot climates have evolved other strategies to minimize the impact of their hostile environment on homeostatic mechanisms. For example, fighting is rare in arid-zone species, which thus avoid heat generation, and many species have long-lasting pair bonds which obviate the need for energy-intensive mate-seeking display activity and courtship. These concepts are discussed and amplified by Davies (1982).

The range of behavioural and physiological adaptations which in sum total ensure the survival of the species has been well illustrated for the Sand Grouse by Thomas (1984). A multiplicity of these adaptations have been listed by this author and include the following.

(i) Selection of appropriate micro-environments—movement between sun and shade is reminiscent of heliothermic reptiles, conserving energy and water reserves.
(ii) Activity (flying and feeding) in the morning and evening, when metabolic heat can be dispersed most easily without invoking evaporative mechanisms.
(iii) Thermal insulation by feather erection and huddling with conspecifics both at low temperature (energy conservation) and when ambient temperatures exceed body temperatures (water conservation).
(iv) Infrequent drinking (in some species at least), allowing exploitation of wider areas around watering points, and saving water and energy on drinking flights.

(v) Reduced metabolic rate and selection of energy- and protein-rich seeds reducing food requirements, metabolic heat loads and possibly foraging time.
(vi) An excretory system apparently well adapted for water and salt conservation.
(vii) Specialized reproductive biology, reducing the metabolic demands for clutch formation and egg-water loss, and allowing the young to be watered without drawing on parental water reserves.

As a general principle, it is suggested that successful desert animals are also likely to show a similar multiplicity of adaptations, whose concerted effect is to conserve water and energy reserves (Thomas, 1984).

5.7 The gut

Reference has already been made to the importance of the gut in osmoregulation. First, it is the interface with the environment, across the mucosa of which water and electrolytes gain access to the internal medium of the body. Also, as we have seen, the cloaca, rectum and caeca have an important role in the final pattern of adjustment of salt and water balance involving also the post-renal modification of urine. It is logical to assume that, given the same absorptive efficiencies, birds with larger guts would have an advantage over birds with smaller guts. In terms of salt and water balance, the surface area of the integrative segment and caeca would be relevant to the overall problem of homeostasis. The acquisition of water and electrolytes is not unconnected with the overall need to provide energy from foodstuffs and there is now evidence which suggests that birds with access to energy-rich sources of food have shorter guts than birds feeding on low-quality food. Moreover, birds such as pigeons, starlings and quail have been shown to alter the gut length depending on the nutritional quality of the available food. These observations are of relevance in strategies for predator avoidance. Longer guts, with the commensurate longer retention of gut contents to improve the digestive efficiency, carry the penalty of greater body weight and resulting diminished manoeuvrability. Direct evidence for this is hard to obtain, but support comes from the observation that hawk kills on wood pigeons are most frequent at dusk at a time when pigeons are at their feeding peaks and their crops are full of food (Sibley, 1981). The crop also permits the rapid intake and storage of water which has advantages for birds who pay episodic visits to water sources.

CHAPTER SIX

THE REPRODUCTIVE SYSTEM AND ITS FUNCTIONS

Unlike some reptiles and most mammals, birds have not adopted the strategy of viviparity (the production of live young) to protect their developing embryos. The embryonic bird can not therefore draw directly on its mother for nutrients, but depends on a large store of yolk. This is laid down in the cell which is destined to become the female gamete while it is enclosed within a follicle in the ovary. This yolk-filled cell is released from the ovary during the process of ovulation and is transported to the outside world by the oviduct. Fertilization must occur in the upper end of the oviduct before the yolk is surrounded by a series of protective layers. These layers are essential for the survival of the developing embryo. One of these is a thick layer of albumin, or 'white', which protects the embryo from desiccation and because of its anti-bacterial properties affords protection against infection. Physical protection for the developing embryo is provided by a calcified shell. Both albumin and shell are produced by specialized regions of the oviduct. Once the fertilized, shelled egg is laid, the fate of the developing embryo within it depends on the parental behaviour of one or both of its parents. They must adopt a strategy to incubate the egg and to turn it regularly to prevent the embryo sticking to the shell. In species with precocial young, the newly-hatched chicks are well feathered with open eyes, and require less parental care than the chicks of species with altricial young which hatch at an earlier stage of development.

The reproductive system of the male bird has fewer characteristically avian features than that of the female. As in reptiles and some mammals, the testes are located in the body cavity. In many species of birds, the copulatory organ in the male is not well developed making the act of internal fertilization a skilful achievement.

In this chapter, we describe the components of the reproductive system and the hormones they secrete and consider how they interact to produce the fertile egg, and finally, the newly-hatched bird.

6.1 The gonads and reproductive tract

The gonads produce male or female gametes and steroid hormones. The steroid hormones stimulate sexual behaviour and the development and functions of the reproductive tracts. Detailed accounts of the structure and function of the female and male reproductive systems are given by Gilbert (1979) and Lake (1981), respectively.

6.1.1 *The ovary*

In most species, the development of the right ovary is suppressed. Notable exceptions are found in certain birds of prey and in the Kiwi. The ovary is thus located on the left in the anterior part of the body cavity, suspended from the dorsal body wall by a peritoneal fold. At hatch the ovary contains millions of oocytes, many of which will degenerate; only a minute proportion will mature and ovulate. Associated with these oocytes, within the connective tissue are interstitial cells which produce steroids. An oocyte which is destined to ovulate must first be incorporated into an ovarian follicle (Figure 6.1). The follicle supports the oocyte, which grows so large during its development that its own fragile cell membrane would otherwise rupture. The follicle also serves to extract yolk material, produced in the

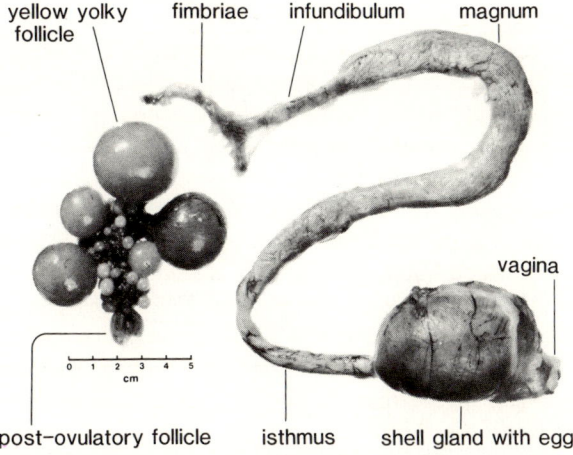

Figure 6.1 The ovary and oviduct of the domestic hen. The yellow-yolky follicles in the ovary are arranged in a hierarchy of sizes, each destined to ovulate on successive days. An egg can be seen in the shell gland.

liver, from the blood and transport it to the growing oocyte. The development of a follicle around the oocyte starts with the formation of a layer of granulosa cells. A second layer of cells of different embryological origin, termed the thecal layer, is subsequently formed around the granulosa layer. Further supportive layers are laid down as the follicle grows, until it begins to extrude from the surface of the ovary. As yolk deposition speeds up, the follicle becomes detached from the main body of the ovary but remains connected to it by a follicular stalk. When the oocyte is fully developed, ovulation occurs as the result of the rupture of the follicular wall. The number of large yellow-yolky follicles which develop in the ovary is related to clutch size. In the domestic hen and other species which have a large clutch size, the yellow-yolky follicles form a hierarchy so that the ovary looks like a bunch of grapes (Figure 6.1).

During the onset of sexual development the growing ovarian follicles and the interstitial cells secrete increasing amounts of oestrogens. These stimulate the synthesis of yolk in the liver and the growth of the oviduct while behaviourally they stimulate sexual interest in the male. Once yellow yolk begins to accumulate, the follicle grows very rapidly and the total weight of the ovary can increase 100-fold or more. As the yellow-yolky follicle enlarges, it produces increasing amounts of oestrogen, originating in the thecal cells. But as follicular growth continues, and ovulation approaches, the production of oestrogen by the thecal cells falls, while that of progesterone by the granulosa cells increases (Figure 6.2). The increasing

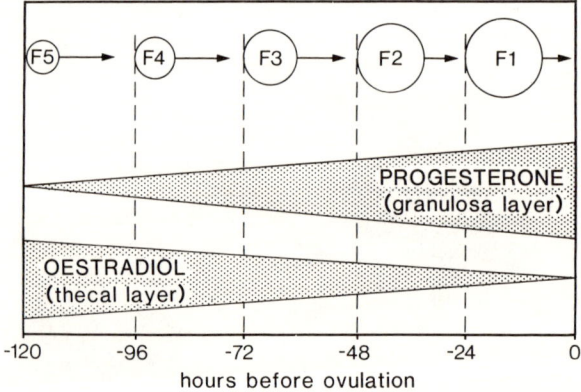

Figure 6.2 A schematic diagram showing the changes which occur in the concentrations of steroids in the granulosa and thecal layers of rapidly growing yellow-yolky ovarian follicles ($F_5 - F_1$). These changes are directly related to the quantities of these steroids released into the blood. (Based on Bahr *et al.*, 1983.)

concentrations of plasma oestrogen coming from the smaller yellow-yolky follicles, intensifies courtship behaviour, stimulates further development of the oviduct and mobilizes calcium reserves in the bones in preparation for eggshell formation. As ovulation approaches, progesterone secreted by the largest yellow-yolky follicles promotes the final development of the oviduct and induces copulatory and nest-building behaviour.

6.1.2 The oviduct

Only the left oviduct normally develops, and is held in place in the body cavity by dorsal and ventral ligaments. The size of the oviduct depends on the secretion of oestrogen and progesterone by the ovary: it enlarges only when the bird is coming into breeding condition (Figure 6.1). The internal opening of the oviduct, the infundibulum, is surrounded by ciliated finger-like processes called fimbriae, which during ovulation are engorged with blood and are highly motile. They engulf the ovulating follicle, thus ensuring the safe passage of the oocyte into the oviduct. After fertilization, the first layer of albumin is laid around the yolky zygote while it is still in the infundibulum. In the chicken, the egg remains in the infundibulum for 15–30 min before passing to the longest part of the oviduct, the magnum. Here most of the albumin is deposited, a process which requires 2–3 hours. The egg then passes to a shorter section of the oviduct, the isthmus, where over a period of about 1–5 hours, membranes are laid down giving the egg its typical shape. The egg then enters the shell gland (or uterus) where it remains for about 20 hours. During the first six hours or so, water passes into the egg in a process known as 'plumping' and is taken up by the

Table 6.1 Mean duration of the fertile period in some domesticated and wild birds (from Lake, 1975 and unpublished data)

Species	Mean duration of fertile period (days)
Domestic fowl	12
Domestic goose	8
Duck	7
Quail	6
Domestic turkey	28
Ring Dove	8
Ring-necked Pheasant	18
Guinea Fowl	7
Northern Fulmar	21
Grey-faced Petrel	60

albumin, resulting in a doubling of its mass. Thereafter the outer membrane becomes progressively calcified, forming a hard shell. Finally, shortly before laying, pigments and a cuticle are deposited on the shell. The vagina serves to transport the egg from the shell gland to the cloaca and is not involved in egg formation. The vagina, however, has another role to play. At its junction with the shell gland there are specialized tubular pockets known as sperm host glands. These store spermatozoa and enable the bird to produce fertile eggs for many days after mating (Table 6.1).

6.1.3 *The testes and the male reproductive tract*

The paired testes lie in the body cavity, suspended from the dorsal wall and are filled with a continuous network of convoluted seminiferous tubules. These contain the germ cells and non-germinal Sertoli cells. The germ cells give rise to spermatogonia which in breeding birds undergo several mitotic divisions, forming more spermatogonia or primary spermatocytes. Two meiotic divisions follow, yielding successively secondary spermatocytes and spermatids which then differentiate into mature spermatozoa. The Sertoli cells are probably a local source of steroids and other hormones which contribute to the development of the spermatozoa. The mature spermatozoa leave the testis via a long coiled duct attached to its side, the epididymus, carried in seminal fluid secreted by this structure and by the seminiferous tubules. The *ductus deferens* serves to transport the semen to the cloaca, adding a little fluid of its own on the way. In sexually active birds, the *ductus deferens* is highly convoluted and turgid, being filled with milky-white semen. In some passerine species, the caudal end of each *ductus deferens* forms a knot-like mass of convulutions which project from the surface of the body looking remarkably like testicles within a scrotum. Close to the cloaca, the *ductus deferens* often expands into a receptacle or seminal sac and during copulation these sacs are everted into the cloaca. Avian semen is extremely viscous due to the high density of spermatozoa per unit volume. The volume produced is small (e.g. 80–500 μl in a cockerel, 50–200 μl in a Golden Eagle) and during copulation is deposited directly into the vagina which is everted into the cloaca. A small erectile phallus is found in most species and as in the chicken, consists of a pair of round folds flanking a small central white body. A well developed protrusible phallus occurs in the Ratites (e.g. ostrich, cassowary) and closely related tinamous, and in the Anseriformes (e.g. ducks and geese).

Androgen-secreting Leydig cells lie between the seminiferous tubules. In juveniles or in birds which have finished breeding the cells are small and contain little lipoidal material. Before the onset of breeding stores of cholesterol-rich lipid, the precursors required for the synthesis of androgens, accumulate in the cytoplasm. At the height of the breeding season the cells become depleted of lipids. The principal steroid secreted by the Leydig cells is testosterone but androstenedione and 5α-dihydrotestosterone, which may also originate in the testis, are also important androgens found in the blood of male birds.

6.2 The brain and pituitary gland

Gonadal activity and reproductive behaviour is controlled by the brain (Figure 6.3). Information from the external environment, such as changes in daylength, social conditions, food availability, and from the internal environment, relating to reproductive status or general bodily health, is integrated in the brain to modulate the secretion of gonadotrophin-releasing hormone (GnRH) and prolactin-releasing or inhibiting factors.

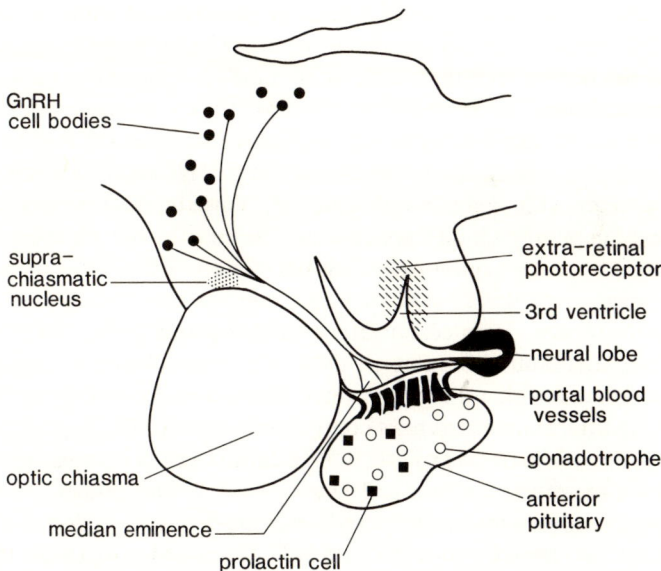

Figure 6.3 A sagittal view of the hypothalamus of a bird to show the relationship between the neuronal cell bodies containing gonadotrophin-releasing hormone (GnRH), the median eminence, the portal blood vascular system and the anterior pituitary gland.

Only small amounts of these neuropeptides or neurotransmitters are released from the brain and they function as neuroendocrine signals which are amplified by the pituitary gland. Thus, relatively large amounts of luteinizing hormone (LH) and follicle-stimulating hormone (FSH) are released from the pituitary gland in response to a small release of GnRH. The concentrations of LH and FSH released into the blood are directly related to gonadal activity and are regulated by the inhibitory or stimulatory feedback effects of gonadal steroids. Gonadal steroids also affect specific regions of the brain to stimulate different components of reproductive behaviour.

6.2.1 *Gonadotrophin-releasing hormone*

Chicken gonadotrophin-releasing hormone (GnRH) occurs in two forms: both are decapeptides having the amino acid sequences pGlu-His-Trp-Ser-Tyr-Gly-Leu-Gln-Pro-Gly-NH_2 (LHRH-I, King and Millar, 1982) and pGlu-His-Trp-Ser-His-Gly-Trp-Tyr-Pro-Gly-NH_2 (LHRH-II, Miyamoto *et al.*, 1984). It is probable that both forms of GnRH occur in other birds. Mammalian GnRH and both forms of chicken GnRH stimulate the release of LH and FSH from the avian pituitary gland.

Nerve cell bodies and fibres containing GnRH are found principally in the hypothalamus. This is the part of the brain which plays a central role in the regulation of reproductive function (Figure 6.3). The GnRH cell bodies send long axons to terminate at the surface of the hypothalamus in an area known as the median eminence (Figure 6.3). This structure is covered in a network of capillary blood vessels which drain into the pituitary gland (Figure 6.3). Gonadotrophin-releasing hormone is released from the median eminence into this blood capillary system to stimulate the release of the gonadotrophins from cells in the pituitary gland.

The hypothalamus contains many neural circuits which come into synaptic contact with GnRH neurones. These are controlled by stimuli originating externally or from different components of the hypothalamic–pituitary–gonadal axis (Sharp, 1983). The neural circuits involved make use of different neurotransmitters and neuropeptides. The neurotransmitters which have been most extensively studied are noradrenaline, dopamine and 5-hydroxytryptamine (Scanes *et al.*, 1982; El Halawani *et al.*, 1982). They all occur in the median eminence and in the region of the GnRH cell bodies. Both are sites at which the release of GnRH may be controlled.

6.2.2 Follicle-stimulating hormone and luteinizing hormone

Follicle-stimulating hormone (FSH) and luteinizing hormone (LH) have been purified from the pituitary glands of the chicken, turkey and ostrich, and have separate biological actions on the gonads (Goldsmith and Follett, 1980; Bono-Gallo et al., 1983). Follicle-stimulating hormone stimulates the growth of the gonads and certain aspects of steroidogenesis while LH stimulates the release of gonadal steroids. The growth of the ovary (Etches and Cheng, 1981) and of the testis (Ishii, 1980) is thus associated with a marked increase in FSH receptors. Luteinizing hormone acts specifically on the Leydig cells to release testosterone, and on the granulosa cells from mature ovarian follicles, to release progesterone (Figure 6.4). In endocrine terms, the onset of breeding is initiated by an increase in the secretion of FSH and LH from the pituitary gland. The secretion of gonadotrophin secretion is rarely constant but occurs in a series of pulses (Figure 6.5). The frequency and the height of these pulses determine the overall concentrations of gonadotrophins in the blood. Similar pulses can be induced by injections of GnRH and are therefore thought to be due to pulsatile releases of GnRH from the median eminence. The rate of secretion of gonadotrophins may also change during the day. For example, in juvenile male and female chickens concentrations of plasma LH are higher at night than during the day (Sharp, 1983).

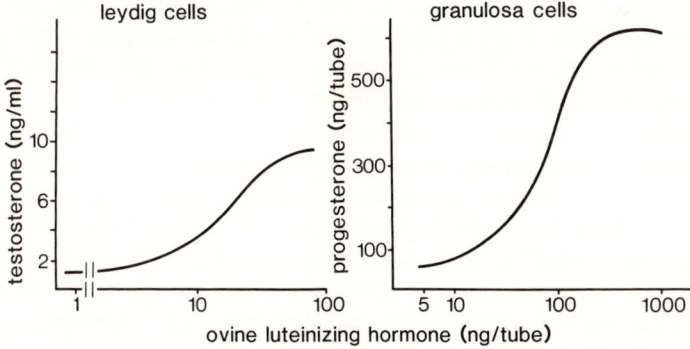

Figure 6.4 The stimulatory effects of ovine LH on the release of testosterone from quail testes Leydig cells and of the accumulation of progesterone in hen ovarian granulosa cells. The Leydig or granulosa cells were isolated and incubated with LH for 3 hours. (Leydig cells: Maung and Follett, 1977; granulosa cells: Marrone and Hertelendy, 1983.)

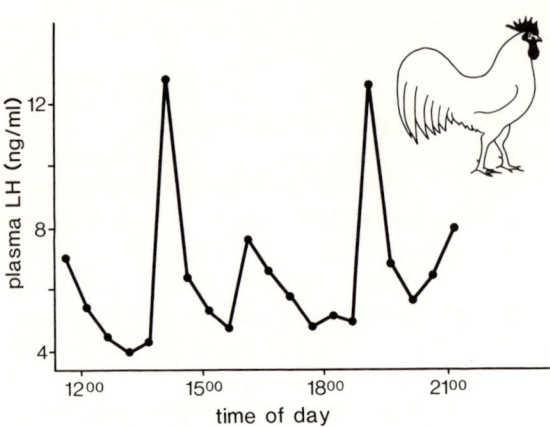

Figure 6.5 Pulsatile release of LH in an adult cockerel. Each pulse is due to a rapid discharge of GnRH from the hypothalamus. (From Sharp, 1983.)

6.2.3 *Inhibitory steroid feedback*

The concentrations of plasma gonadotrophins and gonadal steroids are maintained in a state of dynamic equilibrium in juvenile and adult birds. In order to demonstrate this feedback relationship it is necessary to show that gonadotrophins stimulate the secretion of steroids which, if removed, cause an increase in gonadotrophin secretion. We have already noted in the male that LH stimulates the secretion of testosterone from the Leydig cells and in the female, LH stimulates the secretion of progestesone and oestrogens. If the concentrations of these steroids in the blood are lowered by gonadectomy, concentrations of plasma gonadotrophins increase rapidly. This is achieved by an increased frequency of pulsatile discharges of gonadotrophins. Treatment of the gonadectomized bird with gonadal steroids reverses these changes, causing concentrations of plasma gonadotrophins to fall (Figure 6.6). Steroids from sources other than the gonads, such as the adrenals or from the diet, can also inhibit gonadotrophin release, but they do not form part of a feedback system. In addition to testosterone and oestrogens, other steroids are also involved in the inhibitory feedback control of gonadotrophin release including androstenedione in the male and progesterone in the female.

Steroids exert their inhibitory effects on gonadotrophin release by decreasing the response of the pituitary gland to GnRH and by directly inhibiting the release of GnRH. Oestrogens are the most potent steroid inhibitors of gonadotrophin release and are particularly effective at the

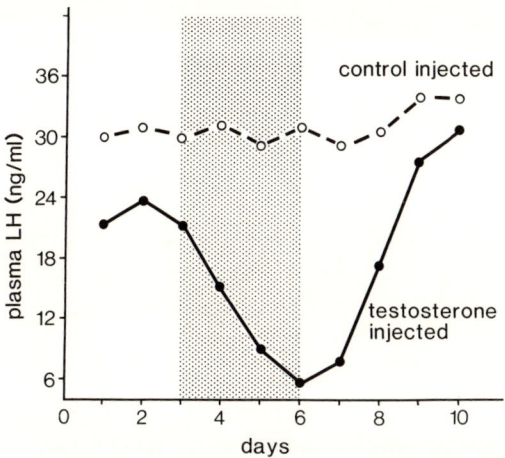

Figure 6.6 The inhibitory feedback effect of testosterone on LH secretion in a castrated cockerel. The removal of the testes has resulted in very high concentrations of plasma LH. Treatment with 1 mg of testosterone per day for 3 days (shaded area) depressed LH concentrations to the values seen in intact cockerels. (P.J. Sharp, unpublished data.)

level of the pituitary gland. In addition to acting on the pituitary gland, testosterone inhibits the release of GnRH from the hypothalamus (Sharp and Gow, 1983). This could be an indirect effect, since testosterone is metabolized locally in the hypothalamus to other inhibitory feedback steroids including oestrogens (Sharp, 1983).

6.2.4 *Stimulatory steroid feedback*

Ovulation is induced by LH, and consequently during the pre-ovulatory period, LH release must temporarily escape from inhibitory feedback control. We have already seen that as an ovarian follicle approaches ovulation the chief steroid it produces is progesterone, in the granulosa layer (Figure 6.2). Furthermore, the release of progesterone from the granulosa layer is readily stimulated by LH. By some unknown mechanism (see section 6.5.3), progesterone from the pre-ovulatory follicle stimulates the release of LH which stimulates further progesterone release. In this way a cascade mechanism is triggered, building up pre-ovulatory releases of LH and progesterone. The stimulatory feedback action of progesterone is mediated by the hypothalamus, resulting in an increase in the release of GnRH. Thus, if this release of GnRH is neutralized by anti-LHRH serum,

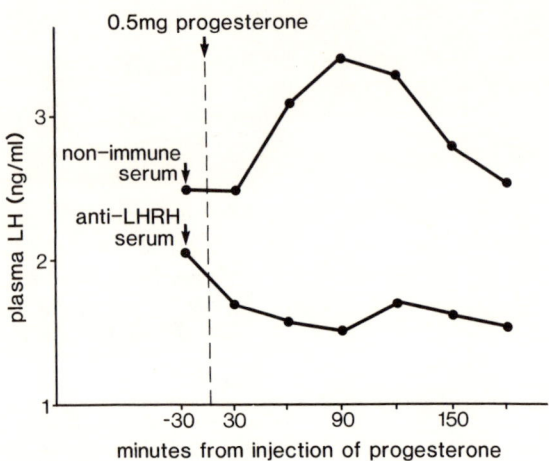

Figure 6.7 An experiment in laying hens showing that the stimulatory effect of progesterone on LH release depends on the release of GnRH. A pre-ovulatory-like release of LH induced by an injection of progesterone is abolished by prior treatment with anti-GnRH serum. It is concluded that progesterone exerts its stimulatory feedback effect by stimulating the release of GnRH. (From Fraser and Sharp, 1978.)

the stimulatory feedback action of progesterone on LH release is abolished (Figure 6.7).

The stimulatory feedback action of progesterone on LH release depends on the presence of oestrogen. Thus, progesterone does not stimulate LH release in out-of-lay or ovariectomized hens. However, if an ovariectomized hens is first 'primed' with injections of oestrogen, progesterone then stimulates LH release (Sharp, 1980).

6.2.5 Prolactin and its central control

Prolactin has been purified from the pituitary glands of chickens and turkeys (Goldsmith and Follett, 1980) and amongst many other functions it is involved in the control of reproduction (Goldsmith, 1983). It is sometimes known as the 'maternal hormone', since increased prolactin secretion is associated with incubation behaviour. In pigeons and doves it stimulates the hypertrophy of the crop sac, resulting in the production of curd-like 'crop milk' which is used to feed newly-hatched chicks. Many species develop a highly vascularized, defeathered brood patch which is used to maintain the egg at body temperature during incubation. The formation of this structure is also partly dependent on prolactin, together with ovarian steroids (Drent, 1975). High concentrations of prolactin in the blood are often associated with low levels of plasma gonadotrophins and in some species, injections of prolactin induce gonadal regression. Prolactin is therefore often regarded as being 'anti-gonadotrophic'.

Like FSH and LH, the secretion of prolactin from the pituitary gland is controlled by the brain. The avian hypothalamus contains both prolactin-releasing and -inhibiting factors, but the prolactin releasing factor is present in larger amounts (Chadwick and Hall, 1983). The identity of prolactin-releasing factor is uncertain, but the prolactin-inhibiting factor is dopamine.

6.2.6 Gonadal steroids and sexual behaviour

Sexual behaviour is dependent on the presence of gonadal steroids—oestrogen and progresterone in the female—and testosterone in the male. Gonadectomy results in the disappearance of sexual behaviour, while the administration of gonadal steroids causes its expression. Treatment of gonadectomized birds with different steroids has led to the discovery that specific components of sexual behaviour are dependent on particular steroids or combinations of steroids. For example, in ovariectomized Ring Doves, oestrogen stimulates courtship behaviour while the addition of progesterone to the oestrogen treatment facilitates nest-building and incubation in response to the presence of nest material and eggs, respectively (Cheng and Silver, 1975). In the male, different components of sexual behaviour are facilitated by direct and indirect actions of testosterone. The indirect effects of testosterone are mediated by metabolites of the steroid produced locally within the brain. Thus, testosterone can be aromatized in the brain to oestrogen or reduced to 5α- or 5β- metabolites. For example, in male quail neural aromatization of testosterone to oestrogen is important for the expression of copulatory behaviour (Adkins et al., 1980) whereas the 5α- reduced metabolites of testosterone activate strutting and crowing behaviour (Adkins and Pniewski, 1978). In male Ring Doves, aggressive components of courtship behaviour depend on testosterone and its 5α- reduced metabolites while the non-aggressive nest-orientated components of courtship depend, in part, on the aromatization of testosterone (Adkins-Regan, 1981). In chickens and quail, female sexual behaviour can be readily elicited in either sex by oestrogen. However, the converse does not apply: male sexual behaviour, and in particular, male copulatory behaviour, cannot readily be induced in females by testosterone. The areas of the avian brain which control male copulatory behaviour are thus sexually differentiated (Adkins, 1978). The process of sexual differentiation depends on the presence of gonadal steroids during a 'critical period' of embryogenesis. The male is the neutral sex in as much as male copulatory behaviour is expressed in the adult provided that the embryo is not exposed to oestrogen or testosterone during the critical period. This critical period occurs before days 12 and 13 of incubation in quail and chickens respectively. Exposure of embryonic chickens and quail to oestrogen or testosterone during this critical period therefore, effectively feminizes or more accurately, demasculinizes, their sexual behaviour when adult. The areas of the brain controlling mating behaviour have been located by using small implants of testosterone, lesioning and autoradiography of injected androgens. The principal area concerned with this behaviour is located in the preoptic area of the hypothalamus (Barfield, 1971; Meyer, 1973).

Another aspect of sexual behaviour, male territorial and display singing, also depends on gonadal steroids. Thus, this type of singing is abolished in Canaries and Zebra Finches after castration and is reinstated after treatment with testosterone. This behaviour is sexually dimorphic in the Zebra Finch since testosterone will not induce song in the female. The development of the morphological and functional capacity for male song depends on early exposure to oestrogen (Gurney and Konishi, 1980).

The morphological structures concerned with male song include a series of brain nuclei from the telecephalon to the medulla. These song control nuclei are markedly sexually dimorphic, being much larger in sexually mature males than in females. Two of these telencephalic song control nuclei in the male Canary, the hyperstriatum ventrale pars caudale (HVc) and the robust nucleus of the archistriatum (RA) are, respectively 99 and 76 per cent larger in the spring, when the birds are singing, than in the autumn at the end of the breeding season (Nottebohm, 1981). The seasonal changes in the size of the HVc and RA are androgen-dependent and are due to the generation, growth and subsequent degeneration of neurones which are specifically concerned with song production (DeVoogd and Nottebohm, 1981; Paton and Nottebohm, 1984).

6.3 Interactions with thyroid hormones

Reproductive function is strongly influenced by thyroid hormones. In several species, including chicken, quail, duck and Weaver Finch, thyroidectomy inhibits the growth of the gonads. This effect can be reversed by giving thyroid hormones. However, the dose is critical: very high doses can inhibit reproductive function. In some species, notably the Munia finches, thyroid activity is sufficiently high in the non-breeding bird to inhibit gonadal development. Thus, if the thyroid glands are removed, the birds come into breeding condition but the gonads can be made to regress again by giving thyroxine (Thapliyal and Chandola, 1972). Once in breeding condition thyroidectomy prevents the normal seasonal regression of the gonads. This effect of thyroidectomy is not restricted to the Munia finches: in the starling too, thyroidectomy prevents seasonal testicular regression (Wieselthier and van Tienhoven, 1972).

These effects of thyroidectomy and thyroid hormone administration suggest one of the ways in which changes in the bird's environment may influence reproductive function. The concentrations of thyroid hormones in the blood are regulated by food intake which depends on food availability and energy requirements, and by other environmental factors such as temperature and daylength (Sharp and Klandorf, 1984).

6.4 Interactions with the external environment

Outside the equatorial regions of the world, the development of the reproductive system is often related to seasonal changes in daylength (see Chapter 7). If a bird is to use daylength to regulate reproductive function it requires a photoreceptor to detect light and a biological clock to measure how many hours of light there are in a day. A photoreceptor and biological clock are also required for the timing of the pre-ovulatory release of LH (section 6.5).

6.4.1 *Photoreceptors*

The most obvious photoreceptor is the eye. However, gonadal growth and regression occurs in response to changes in daylength even after blinding in many photoperiodic species (Follett, 1984). The major photoreceptor for photoinduced gonadal growth must therefore be located elsewhere. The pineal gland located on the dorsal surface of the brain and sometimes referred to as 'the third eye' is another probability. But again, studies on pinealectomized birds show that it is not a major photoreceptor for gonadal growth (Follett, 1984). The most likely site of the photoreceptor is in the basal hypothalamus, just above the median eminence (Figure 6.3). If this area is illuminated using optic fibres or minute pellets of radioluminous paint, gonadal growth is stimulated in birds (e.g. quail, White-crowned Sparrow) kept on a non-stimulatory daylength. Illumination of other areas of the brain in a similar manner does not stimulate gonadal activity. The structures in the basal hypothalamus concerned with photoreception do not have any obvious morphological characteristics. However, they are more sensitive to red than to green light.

6.4.2 *The biological clock*

The biological clock is not a single structure but is composed of several clocks in an interlocking system (Menaker *et al.*, 1981). The pineal gland, the suprachiasmatic nucleus, (see Figure 6.3) and perhaps the retina of the eye are capable of generating circadian rhythms of neural or endocrine activity. If the pineal gland of the chicken is maintained in culture in constant darkness, it releases melatonin rhythmically, with peaks of secretion occurring approximately every 24 h. Evidence that the suprachiasmatic nucleus is a biological clock comes from studies on birds in which this structure is destroyed by small electrolytic lesions. In such birds, the free-running activity rhythms with a period of about 24 h which are normally observed in constant lighting conditions are abolished. In view of the complexity of the bird's biological clock, it is not surprising that much is still to be learned about the way in which it is involved in the regulation of photoinduced gonadotrophin secretion, and in the control of ovulation.

6.5 The ovulatory cycle

Some birds (e.g. shearwaters and petrels) produce only one egg; others lay many (e.g. game birds, ducks, passerines), often on successive days, but sometimes separated by longer intervals. The ecological factors determining clutch size are well understood (Perrins and Birkhead, 1983) but knowledge of the physiological processes involved is restricted to a few domesticated species. The ovulatory cycle of birds differs from the oestrous cycle of mammals in that the post-ovulatory follicle does not form a corpus luteum and, therefore, there is no luteal phase.

6.5.1 *The timing of ovulation*

It is generally observed that birds lay their eggs during the same period each day and that this period differs between species. For example, ducks lay early in the morning, chickens in the late morning and turkeys and quail in the afternoon. These patterns of lay reflect the fact that ovulations and, hence, pre-ovulatory releases of LH occur during a restricted period of the day only (Sharp, 1983). The period of the day during which pre-ovulatory releases of LH may be initiated is known as the 'open period' of the ovulatory cycle. In the domestic hen, it lasts between 8 and 10 h. Pre-ovulatory releases of LH occur in an orderly sequence within the 'open period' on successive days. Thus, LH release is first initiated at the beginning of the 'open period' with the resulting egg being laid early in the day. Pre-ovulatory releases of LH occur a little later within the 'open period' on successive days with corresponding delays in the times at which the resulting eggs are laid. Finally, at the end of the 'open period', a pre-ovulatory release of LH cannot be initiated and on that day there is no ovulation. This pattern of pre-ovulatory release of LH results in a characteristic pattern of lay, with a series of eggs being laid on successive days separated by pause days on which no eggs are laid. Each series is termed a 'sequence'. The term 'clutch' refers to the total number of eggs laid before the onset of incubation and could contain two or more sequences. The length of a sequence depends on the rate at which the yellow-yolky follicles develop in the ovary. If the follicles mature at a rate of one every 24 h then sequences can be very long: some hens can lay 100 or more eggs on successive days. If the follicles grow more slowly, then the cycle of follicular maturation is asynchronous with the 24 h rhythm underlying the

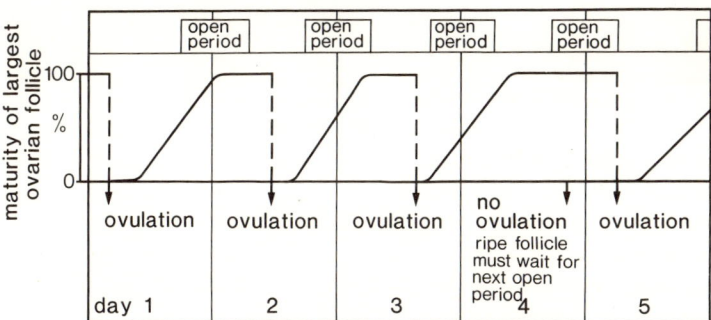

Figure 6.8 Diagram to explain the laying of eggs in sequences separated by 'pause days' in the domestic hen. The 'open period', during which pre-ovulatory releases of LH are initiated, occurs every 24 hours. Yellow-yolky follicles capable of ovulating develop in the ovary about every 26 hours. The resulting asynchrony between the 24 hour 'open period' cycle and the c.-26 h cycle of follicular maturation results in days in which ovulation cannot occur. This in turn results in a day on which no egg is laid. (From Sharp, 1983.)

'open period' (Figure 6.8). Thus, successive follicles become mature later and later in the 'open period' resulting in the progressive delay in the initiation of the pre-ovulatory releases of LH described above.

6.5.2 Circadian rhythms and the 'open period'

When hens are exposed to 24 h daylengths (e.g. 14 h light per day), the 'open period' occurs at night. This observation has lead to the erroneous conclusion that pre-ovulatory LH release is initiated by darkness. In fact, the 'open period' is controlled by the bird's circadian system (see section 7.3.2). The phase of a circadian rhythm in relation to its entraining light–dark cycle depends both on the light:dark ratio and on the period of the lighting cycle. It can be predicted that the phase of the 'open period' should be advanced in hens entrained by lighting cycles of more than 24 h and delayed in birds entrained by cycles of less than 24 h. Experimentally this prediction has been found to be correct. In hens held on a 21-hour lighting cycle, the 'open period', as indicated by the times at which pre-ovulatory LH release are initiated, is delayed, whereas in hens held on a 30-hour lighting cycle, it is advanced (Figure 6.9).

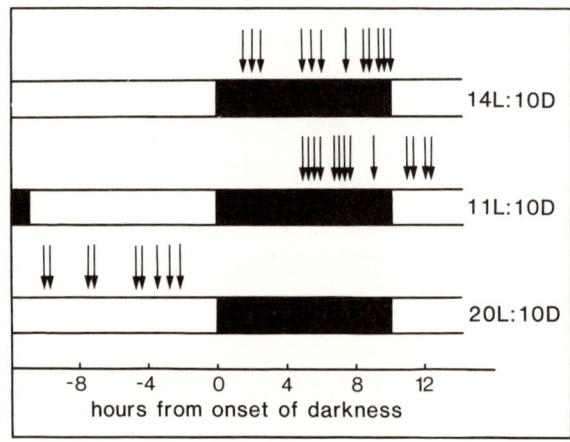

Figure 6.9 Position of 'open period' of the ovulatory cycle of the domestic hen in relation to the onset of darkness in hens entrained to 24-, 21 and 30-h lighting cycles with 10 h darkness in each. The arrows show the times at which pre-ovulatory releases were initiated in different hens studied at the same time and their distribution defines the duration and position of the 'open period'. (From Sharp, 1980.)

6.5.3 Plasma hormones and ovulation

We have already noted that ovulation is caused by the stimulatory feedback action of progesterone on LH release. Thus, the most prominent features of the ovulatory cycle are pre-ovulatory increases in the concentrations of LH and progesterone (Figure 6.10). Another feature is a fall in basal LH concentrations during the period before pre-ovulatory LH release is initiated which, as will be discussed later, may be significant in relation to the mechanism of the 'open period' (see section 6.5.4). Pre-ovulatory LH levels begin to fall some 3 to 6 h before ovulation. The fall is due essentially to a loss of response of the pituitary gland to GnRH. This responsiveness is not recovered until several hours after ovulation. Baseline concentrations of LH increase to broad plateaux between pre-ovulatory LH peaks (Figure 6.10). Concentrations of plasma FSH are also at their highest during the ovulatory cycle at this time. They may be responsible for the recruitment of new follicles into the yellow-yolky follicular hierarchy (Imai, 1983). Several other hormones have been measured during the ovulatory

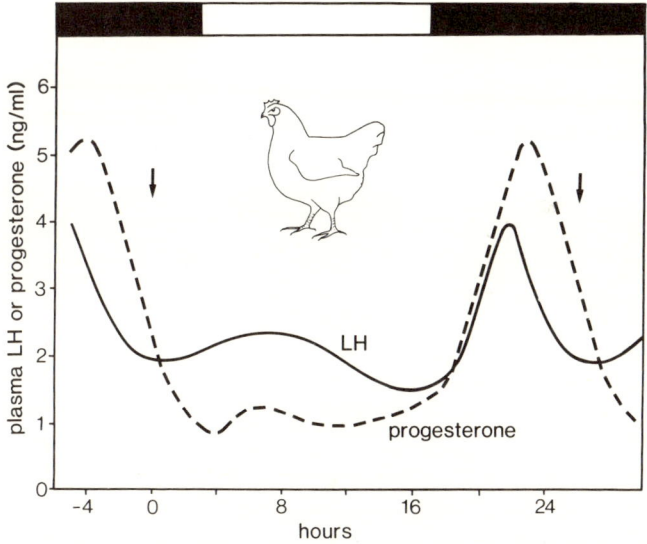

Figure 6.10 Changes in concentrations of plasma LH and progesterone during the ovulatory cycle of hens entrained to a 27-h lighting cycle of 14 h light and 13 h darkness. This lighting cycle keeps pace with the cycle of ovarian follicular maturation so that pre-ovulatory surges are always initiated at the onset of the dark period. The black horizontal bars represent the hours of darkness; arrows show the times of ovulation. (From Sharp, 1983.)

cycle, including oestrogens, androgens and prolactin, but their direct relation to the ovulatory process, if any, has not been clearly established (Sharp, 1980).

6.5.4 Some explanations of the 'open period'

The exact nature of the circadian rhythms involved in the generation of the 'open period' is unknown. Three such rhythms have been proposed and one or a combination of these form the basis of a number of theories. These are:

(1) A circadian rhythm in the responsiveness of the stimulatory feedback system to progesterone.
(2) A circadian rhythm in the basal secretion of LH.
(3) A circadian rhythm in the secretion of adrenal steroids.

(1) As originally proposed by Fraps (1961), during the 'open period' the brain may become more responsive to the stimulatory feedback effect of progesterone on LH release. It is thought that when a follicle is mature, this state of maturity is signalled to the brain by an increase in the release of progesterone into the blood. This is entirely consistent with the observation that follicular maturity is characterized by an increase in progesterone synthesis in the granulosa layer (section 6.1.1). An essential feature of this hypothesis is that there should be a relationship between follicular maturity and the concentration of progesterone in the blood. Since only a 30% increase in plasma progesterone is observed when the ovary contains a mature follicle (Williams and Sharp, 1978) it is not certain that this is sufficient to qualify as Frap's positive feedback signal.

(2) As originally proposed by Bastian and Zarrow (1965) concentration of plasma LH may increase during the 'open period' to induce ovulation, provided a mature follicle is present. Williams and Sharp (1978) modified this hypothesis by suggesting that an increase in baseline LH levels during the ovulatory cycle might stimulate the release of progesterone from a mature follicle and hence trigger the pre-ovulatory cascade of progesterone and LH. This theory can be dismissed, at least in this form, because apart from a small increase in LH secretion observed at the onset of darkness, baseline LH concentrations tend to be depressed during the 'open period'. The possibility that this fall in LH may be important in the control of the 'open period' can also be considered. As shown in Figure 6.10, the pre-ovulatory release in LH is preceded by a fall in LH concentrations. This fall is observed before all pre-ovulatory LH peaks irrespective of where they

occur in a sequence. Wilson and Cunningham (1984) suggest that the fall causes alterations in the receptors for LH in the mature follicle, making it more responsive to LH.

(3) Daily rhythms in concentration of plasma corticosterone coincide with the 'open period' in both entrained and free-running conditions (Wilson and Cunningham, 1984). An increase in plasma corticosterone in combination with the daily decrease in the baseline levels of LH may stimulate the release of progesterone from the mature follicle and hence trigger the pre-ovulatory release of LH by stimulatory feedback.

6.6 Incubation and brooding

Once the egg is laid, the bird must adopt a strategy to incubate it and care for the newly-hatched young (Drent, 1975). Most species maintain the temperature required for incubation by directly transferring heat from their bodies. This may be achieved through the formation of brood patches— highly vascularized areas of skin which lose their feathering just before the onset of incubation. For some species, methods must be found to prevent the eggs from overheating and this can be achieved, for example, by fanning and shading (e.g. ostrich) or by wetting (e.g. some plovers). Brood-parasitism occurs in about 80 species from seven families (e.g. cuckoos, cowbirds, various weaver-finches, honey guides, and one duck). These birds avoid the problems associated with incubation and care of young by laying their eggs in the nests of other species. The Megapodes (e.g. scrub-fowl, brush-turkey) are uniquely distinguished by laying their eggs in a mound of earth or rotting vegetation or, in one species, in volcanic ash, in the warmth of which the eggs are incubated. Some of the Megapodes regulate the temperature of their nest mounds by adding or removing earth or vegetation. After hatching, the young megapode chick struggles to the surface of its nest mound to begin life with no parental contact. Other species, however, brood their young after hatching. Species with altricial young are fed directly by their parents after hatch, notably, by 'crop milk' in pigeons and doves, or by regurgitation of crop or stomach contents (e.g. parrots, gulls, penguins). Other species with altricial young bring food to the nest without prior ingestion. Species with precocious young are generally shown by their parents where and what to eat. Parent birds feeding their young invariably show some form of protective behaviour— shielding the nestlings from adverse climatic conditions and fending off predators.

Incubation and brooding are both aspects of parental behaviour which reflect changes in the functions of the reproductive system and on the concentration of several hormones in the blood. We shall now consider the relationships between these hormones and parental behaviour.

6.6.1 Endocrine changes during incubation and brooding

At the onset of incubation, ovarian follicular growth is inhibited. Any remaining yellow-yolky follicles become atretic and the yolk is resorbed. During the early stages of incubation a few partially developed follicles may persist: these are an insurance against the early loss of the clutch. But as incubation gets under way, the ovaries regress further. The regression of the ovary results in a decrease in the concentrations of plasma ovarian steroids and without the support of these steroids, the oviduct too, regresses. Less information is available about species in which the male participates in incubation, but in the male Ring Dove, testicular function is depressed when eggs are in the nest.

The regression of the reproductive system is caused by a decrease in the secretion of gonadotrophins. Since this is not due to a loss in responsiveness of the pituitary gland to GnRH, it can be concluded that the onset of incubation is caused by a decrease in the release of GnRH.

Perhaps the most interesting change in endocrine function which occurs in incubating birds is a dramatic increase in the secretion of prolactin which is followed, at some time after the chicks hatch, by a steep fall. The precise relationship between these changes in prolactin secretion and the different stages of parental care differ between species and provide clues about the functions of prolactin (Goldsmith, 1983). Three examples are given which may be regarded as typical of species with different parental strategies.

(1) *The Ring Dove: a species producing crop milk.* The obvious function for prolactin in the Ring Dove is that it stimulates the development of the crop sac. The weight of this structure increases rapidly in both sexes towards the end of incubation in preparation for the squabs (nestling doves) which are due to hatch (Figure 6.11). The increase in weight of the crop is preceded by an increase in the concentration of prolactin which remains high while the squabs are being fed 'crop milk'. As the squabs are gradually weaned the concentrations of plasma prolactin in the parents fall and the weight of the crop sac decreases (Figure 6.11). In addition to its role in stimulating the production of crop milk, prolactin also plays a role in the maintenance of

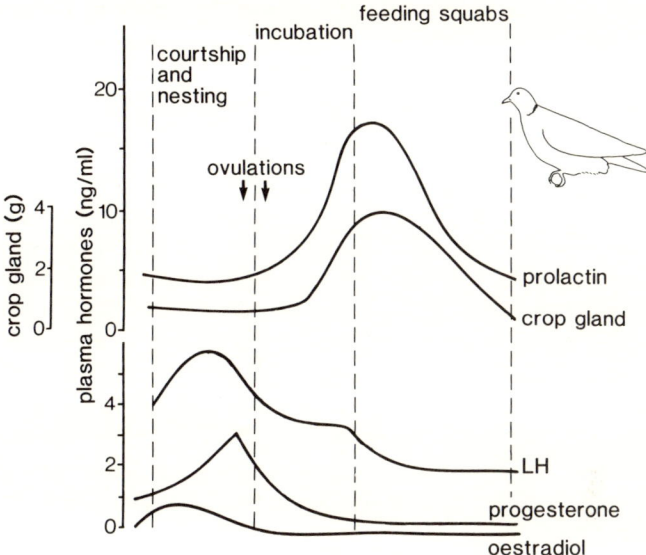

Figure 6.11 Changes in concentrations of plasma hormone levels and crop gland weight during a breeding cycle in the ring dove. (Based on Follett, 1984.)

incubation behaviour. Thus if doves are deprived of their nests at the onset of incubation and are given daily injections of prolactin for several days, they readily resume incubation when their nests are returned. If the nest-deprived birds are given control injections instead, they do not resume incubation when their nests are returned (Silver, 1978).

(2) *The Canary: a non-Columbiform species with altricial young.* As in the Ring Dove, the concentrations of prolactin increase in the Canary during incubation and fall slowly over a period of two or three weeks after hatching. During this period the nestlings are dependent on their parents for food. When the nestlings fledge, prolactin concentrations are low but increase rapidly again at the onset of the next nesting (Figure 6.12). In both the Canary and the dove high concentrations of plasma prolactin are associated with intensive incubation and feeding of the young. One of the probable functions of high concentrations of plasma prolactin at this time is to maintain this behaviour.

(3) *The Turkey: a species with precocial young.* In species with precocial young such as the Turkey, the concentrations of plasma prolactin are also high during incubation. But unlike those in the dove or Canary, they fall

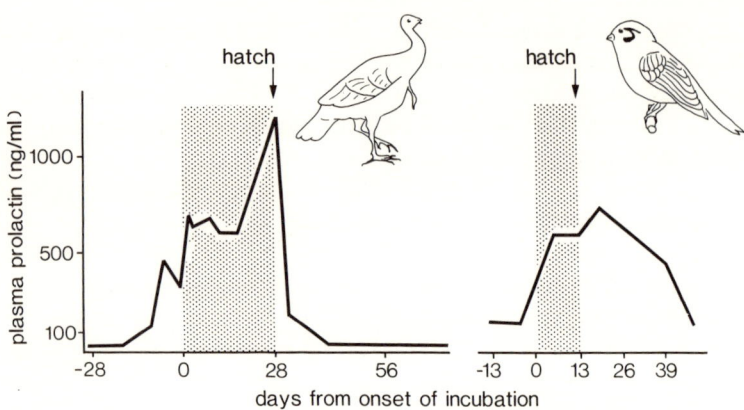

Figure 6.12 Changes in concentrations of plasma prolactin in an altricial species (Turkey) which does not feed its young and a precocious species (Canary) which does. In both species, prolactin levels are high during incubation but remain high after hatch in the Canary but not in the Turkey. (Turkey: Wentworth *et al.*, 1983; Canary: Goldsmith, 1983.)

rapidly when the eggs hatch (Figure 6.12). The intensity of post-hatch parental care in species with precocial young is much less than in species with altricial young and this can be correlated with the concentrations of plasma prolactin. It seems that the parental behaviour directly stimulates the secretion of prolactin, which in turn reinforces the behaviour. Support

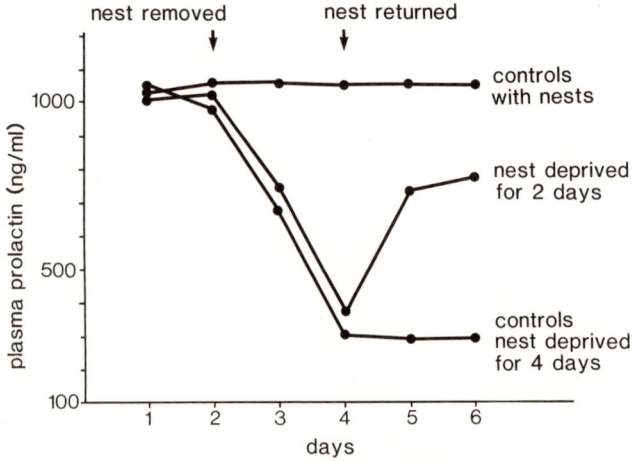

Figure 6.13 An experiment in Turkeys, showing that nesting behaviour stimulates prolaction secretion. Prolactin levels fall after nest removal and increase when the nest is returned. (From El Halawani *et al.*, 1980.)

for this suggestion comes from the finding that concentrations of plasma prolactin fall in incubating turkeys deprived of their nests (Figure 6.13). In this situation, the birds are unable to express the behaviour which is thought to stimulate prolactin release. If the birds are allowed back to their nests, incubation behaviour can again be expressed, resulting in an increase in prolactin secretion (Figure 6.13). Studies on the effects of nest deprivation on concentrations of prolactin have been done in doves and in species with altricial young with similar results: deprivation of the nest, or nestlings, causes a rapid fall in the concentrations of plasma prolactin.

The nest or young stimulates the release of prolactin by at least three pathways: tactile, visual and auditory. The importance of these pathways differs between species. In Ring Doves visual and auditory cues are particularly important, since prolactin secretion can be maintained in an incubating bird if it is is deprived of access but not the sight of its incubating partner. On the other hand, tactile stimuli may be important in species such as the Mallard Duck in which an application of a local anaesthetic to the brood patch causes a decrease in plasma prolactin (Goldsmith, 1983).

We have now identified two roles for prolactin in relation to parental care: crop-sac stimulation and the maintenance of parental behaviour. A further function, referred to previously (section 6.2.5), is an antigonadal action. At the time when concentrations of prolactin are their highest, the reproductive organs have regressed the most. Since the regression of the reproductive system seems to be due to a decrease in the release of GnRH, prolactin may act directly on the brain to suppress this release.

6.6.2 *The initiation of incubation: progesterone or prolactin?*

What causes a bird to stop laying and to begin to incubate? Interest has centred on the possible roles of progesterone and prolactin. Both these hormones stimulate incubation behaviour in male and female Ring Doves. However, the doses of prolactin required to induce incubation are higher than those needed to stimulate the development of the crop sac. Taking this observation together with the fact that the concentrations of plasma prolactin do not increase rapidly in Ring Doves until 4 days *after* the onset of incubation (Figure 6.11), it is concluded that prolactin does not *initiate* incubation. The concentrations of plasma progesterone originating from the pre-ovulatory follicles, however, increase before the onset of incubation (Figure 6.11) supporting a physiological role for this steroid in triggering the behaviour. Concentrations of plasma oestradiol also increase prior to the onset of incubation, but injections of oestrogen do not induce incubation. However, if ovariectomized Ring Doves are treated with oestrogen and/or progesterone, a combination of the two steroids is more effective than progesterone alone in the initiation of incubation behaviour (Cheng and Silver, 1975). It is concluded that in the Ring Dove, progesterone from the pre-ovulatory follicles, supported by oestrogen, together with environmental stimuli from the nest and mate, plays an important role in the initiation of incubation behaviour. Incubation itself causes an increase in prolactin secretion which maintains the behaviour after the concentrations of plasma progesterone and oestrogen decline (Figure 6.11). This explanation for the initiation of incubation applies to the female only since concentrations of plasma progesterone do not increase in the male Ring Dove before incubation (Silver, 1978). In the male, therefore, it seems that the onset of incubation is stimulated by sight of the nest and/or partner sitting on the nest. This, in turn, stimulates the release of prolactin which, as we have seen, maintains incubation behaviour. Since the presence of a nest or incubating female will not induce incubation behaviour in castrated male Ring Doves, the presence of testicular steroids are required for the expression of the behaviour.

In species like the Ring Dove, laying one or two eggs, the transition between laying and incubation is rapid. But in species laying larger clutches such as the bantam hen, the transition can be much longer, with the bird spending progressively more time on its nest each day as the clutch nears completion (Figure 6.14). This progressive development of incubation behaviour may stimulate the gradual increase in concentrations of plasma prolactin before a clutch is completed in several species laying large clutches, including the Turkey (Figure 6.12) and bantam (Figure 6.14). In the bantam hen, this pre-incubation increase in concentrations of plasma prolactin is characterized by pronounced daily rhythms with nocturnal surges which, on successive days before the onset of incubation, spread out into the light period until prolactin concentrations are high at all times (Lea et al., 1982). These nocturnal surges of prolactin are associated with a change in the behaviour of the bird; instead of roosting away from the nest, she now begins to sit on the nest during the night (Lea et al., 1981). It is not known whether these nocturnal surges begin before or after this change in roosting behaviour. An increase in plasma prolactin concentrations *before* the first signs of a switch to incubation behaviour is observed, would support a role for prolactin in the initiation of incubation.

Attempts have also been made to find out whether the initiation of incubation in species with large clutches is caused by an increase in

THE REPRODUCTIVE SYSTEM AND ITS FUNCTIONS

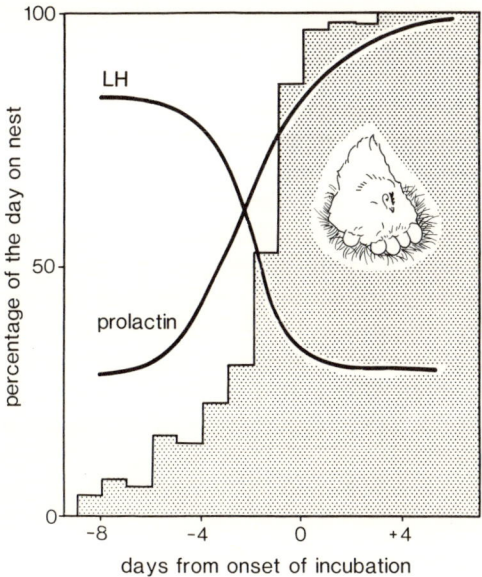

Figure 6.14 Changes in the concentrations of plasma prolactin and LH in bantam hens in relation to the time spent each day on the nest around the onset of incubation. (From Lea *et al.*, 1981.)

prolactin secretion. Despite earlier claims to the contrary, injections of prolactin do not induce incubation in chickens with a history of broody behaviour even though the injections do induce ovarian regression (Table 6.2). In the Budgerigar, however, injections of prolactin stimulate incubation behaviour in the presence of oestrogen (Hutchison, 1975).

Table 6.2 Effects of injections of ovine prolactin on nesting behaviour and ovarian weight in laying hens from a strain prone to broodiness (from Opel and Proudman, 1980)

Treatment[1]	No. of hens	Nest entries per day	Time spent on nest (min)	Ovarian weight (g)
Saline control	10	9	100	33
10 I.U. prolactin	6	6*	66*	20*
20 I.U. prolactin	5	6*	68*	13*
60 I.U. prolactin	6	6*	68*	4*

*Significantly different from controls
[1] Prolactin was injected daily for 6 days and the behaviour was observed for 12 days after the first injection. The ovaries were weighed after 12 days of injections. I.U., international units

In species laying long sequences, the concentrations of plasma progesterone, originating in the pre-ovulatory follicles, are high for many days prior to the completion of the clutch. In this context, it is inappropriate for progesterone to have a rapid effect on the initiation of incubation, otherwise the clutch would be terminated prematurely. Such is not the case in species like the Ring Dove laying only one or two eggs. Not surprisingly, injections of progesterone do not induce incubation in species, like the domestic hen, which have large clutches. Thus, the relative importance of progesterone and prolactin in the induction of incubation in these birds is not known. Taking all the evidence into account it is likely that incubation behaviour is initiated by environmental cues provided by the nest and eggs and that ovarian steroids play an essential, permissive role. Prolactin secretion may also increase as a result of a stimulatory effect of oestrogen on the pituitary gland (Chadwick and Hall, 1983) and could play a role in this process.

6.7 Moult

The end of a breeding season is often terminated by a general body moult. Like reproduction, moult is an energy-demanding process and whenever possible birds partition these two processes within a breeding cycle to avoid energy demands which they might not be able to meet (Payne, 1972). A bird depends on its feathers being in good condition for its immediate survival. So, irrespective of whether it breeds, its feathers must be regularly replaced. There is thus no reason to suppose that the termination of breeding creates the physiological conditions essential for moult. Indeed, some species (e.g. Purple Sandpiper) moult and breed simultaneously.

Thyroid hormones exert a major influence on feather growth. Although feather follicles can form in the absence of the thyroid gland, thyroidectomy interferes with their subsequent growth and development. Thyroid hormones stimulate feather growth, and after administration of these hormones, the newly growing feathers push the old feathers out, thereby causing a moult. The sensitivity of the feather papilla to thyroid hormones is directly related to the time elapsed since the last moult. Thus, moult can be more readily induced with injections of thyroid hormones in birds with old feathers than in birds with new feathers. The underlying factor governing the timing of moult may therefore be a seasonal change in the responsiveness of feather follicles to thyroid hormones.

Several other hormones may also influence the timing of moult. In particular, gonadal steroids inhibit feather growth. In many birds, the administration of androgens or oestrogens slows or inhibits normal moult while feathers lost during the breeding season are replaced slowly or do not regrow (Payne, 1972). Thus, when a breeding season ends, and the concentrations of gonadal steroids fall, the feather follicles are particularly sensitive to the moult-inducing effects of thyroid hormones. This is especially true if the previous major body moult occurred at the end of the previous breeding season. The post-breeding fall in gonadal steroids is often associated with increasing concentrations of thyroid hormones, thereby contributing to the hormonal environment most likely to induce moult.

CHAPTER SEVEN

THE ENVIRONMENT AND REPRODUCTION

The functions of the organs and endocrine systems involved in the efficient production of live young (Chapter 6) have no significance for the survival of the species unless considered in relation to the bird's environment. Natural selection favours a breeding strategy which, in a given environment, is the most likely to ensure (1) the production of the largest number of young which survive to breed, and (2), the survival of parents until they breed again. Since there is no avian equivalent of embryonic diapause or delayed implantation, birds must produce eggs and rear young without interruption. This means that food must be available within a continuous period to support both the increased energy demands of the parents and the growth of the chicks.

Within these constraints, birds have developed an amazing variety of breeding strategies. A study of the *ultimate* factors affecting the evolution of these strategies is the concern of ecologists (Lofts and Murton, 1968; Immelman, 1971; Murton and Westwood, 1977; Perrins and Birkhead, 1983). For many birds, the availability and quality of food required for successful breeding is the most important 'ultimate factor' involved in the timing of breeding. It has survival value for the species without necessarily directly causing breeding. As physiologists, we are concerned with the ways in which birds use environmental information to ensure that they breed at the most appropriate time. Environmental information or 'factors' which directly affect reproductive function are termed *proximate factors*. They can act either as *proximate inductive factors* to stimulate breeding directly or as *proximate phasing factors* to entrain autonomous rhythms of reproductive activity. These two actions of proximate factors are not necessarily mutually exclusive. Proximate inductive factors can be classified in relation to the functions they serve at different stages of the breeding cycle. In this chapter, we consider the diversity of breeding strategies adopted by birds and attempt to identify some of the proximate factors and physiological mechanisms which make them possible.

7.1 Breeding strategies

Some birds live in a constant environment where the food supply does not vary, but the majority live in a non-uniform environment and must breed when the conditions are most favourable. An option taken by some species is to migrate to areas where the food supply is temporarily abundant, and reproduce there (Perrins and Birkhead, 1983). Another, adopted by long-lived species including sea-birds, Ratites (flightless birds) and hawks, is to delay the onset of sexual maturity for one or more years. This strategy is probably related to a decrease in the survival of birds attempting to breed too early.

7.1.1 Migration

Migration occurs twice a year: to the breeding grounds in spring and back again to the wintering quarters in autumn. Migratory behaviour has probably evolved independently in several avian families and its physiological basis may therefore differ between species. The onset of migratory behaviour is characterized by the deposition of body fat and the development of nocturnal restlessness, known as 'Zugenruhe' (see Chapter 3). Changes in climatic conditions and food supply act as proximate factors to trigger migratory behaviour in partial migrants (e.g. European Blackbird) but have little effect on typical migrants (e.g. Willow Warbler) (Berthold, 1975). Experimentally, nocturnal restlessness can be induced or enhanced in partial migrants in autumn by lowering the temperature or restricting food, while the reverse effects are obtained by increasing the temperature and supply of food. In spring or late winter, migratory restlessness and fat deposition can be induced by increasing the temperature. Typical migrants, however, show migratory restlessness irrespective of the changes in environmental temperature or food supply.

Changes in daylength play an important proximal role in stimulating migratory restlessness and fat deposition in many species, although it is not always clear whether they act as inductive or phasing factors or as both. In classical experiments on Slate-coloured Juncos, Rowan (1931) demonstrated that spring-like migratory behaviour could be induced in winter by artificially increasing the daylength. The generality of this observation has been confirmed in several species, notably the Indigo Bunting and the White-crowned Sparrow (Berthold, 1975). Rowan suggested that the effects of changing daylength on pre-migratory fattening and migratory behaviour are mediated by the development of gonads. Thus, the behaviour is thought to be directly dependent on changes in the secretion of gonadal steroids. If this proposition is correct, then migratory behaviour should be abolished after castration. Castration of several species (e.g. White-throated Sparrow) abolishes photo-induced fattening and migratory restlessness, but only if the testes are removed *before* but *not after* the onset of photostimulation. This observation shows that the testes play a role in initiating or phasing the preliminary stages in the complex internal changes which culminate in migratory behaviour. In addition to gonadal steroids, prolactin is also known to play a role in the initiation of migratory behaviour. Injections of prolactin in the White-crowned Sparrow induce migratory restlessness and fat deposition (Meier and Ferrell, 1978). This observation is consistent with the finding that concentrations of pituitary prolactin are at their highest during migration. It seems that the environmental factors which are used by birds to time migration, can either phase autonomous rhythms of migratory behaviour (section 7.4.2) or phase circadian rhythms in such a way that they induce migratory behaviour (section 7.5.3).

7.1.2 *Delayed sexual maturation*

Delayed sexual maturation is related to a failure to secrete gonadotrophins and/or gonadal steroids. For example, in Herring Gulls concentrations of plasma LH levels are directly related to the year class and do not increase during the breeding season until the fourth or fifth year of life (Scanes *et al.*, 1974). Spermatogenesis may occur during the breeding season in sub-adult year classes of several species. The failure to breed is related to reduced or delayed steroid secretion. For example, in second-year (sub-adult) Bar-headed Geese, the seasonal increase in androgen secretion is delayed, making it impossible to establish the behaviour necessary for successful breeding (Dittami, 1981).

7.1.3 *Non-annual breeding*

(a) *Continuous breeding.* It is unusual for wild birds to produce more than 4–5 successive broods, even in an unchanging environment (Immelman, 1971). As we discussed in section 6.7, this is because the energetic demands of reproduction and moult make it impossible in many species for both processes to occur together. Consequently, periods of reproduction alternate with periods of moult. In species which breed throughout the year, individuals or pairs have cycles of reproductive activity which are not synchronous with the rest of the population. Breeding of this type is most commonly found in equatorial rain forests and in some sea-birds (Immelman, 1971).

(*b*) *Breeding cycles of more or less than one year.* In a constant environment, where continuous breeding is possible, excessive predation can be reduced by synchronizing individual breeding cycles within a population. As a result, so many young appear at the same time that their chance of being taken by predators is greatly reduced. Such synchronization is achieved by social interactions within the population. This results in breeding cycles of less than one year, with the breeding period of the whole population shifting from year to year. The best known examples of this type of breeding strategy are provided by sea-birds breeding on tropical islands, the most famous of which is the Sooty Tern of Ascension Island. This species has a breeding cycle of 9.6 months.

If the period of parental care is very extensive, as is the case in the albatrosses, eagles, vultures and the King Penguin, breeding is inhibited during the following breeding season. The physiological mechanism is unknown but a consideration of the relationship between prolactin and parental behaviour (section 6.6.1) suggests that an increased secretion of this hormone could play a role.

(*c*) *Opportunistic breeders.* Opportunistic breeding occurs in environments where changes in the food supply are erratic and difficult to predict. The distinction between opportunistic and seasonal breeding is not always easy to make. Even in the most unpredictable of environments such as the interior of Australia, viewed over a number of years, the frequency of breeding is not the same in each month of the year. A classical opportunistic breeder, the budgerigar, is most likely to breed in the spring, summer or autumn but rarely in the winter in the middle of its range in central Australia. But in the north of its range it breeds in the autumn, and in the south, in the spring (Wyndham, 1980). Similarly, the Zebra Finch, which breeds after rainfall in inland Australia in all months of the year, is more likely to do so in Western Australia in spring when temperatures are rising than in the autumn when temperatures are falling (Davies, 1979).

Within these constraints of underlying seasonality, opportunistic breeders show several physiological adaptations which enable them to take advantage of a sudden increase in food supply and reproduce. A primary adaptation is to maintain a capacity to remain near to breeding condition for a large segment of the year. One curious adaptation of this sort is found in the Pink-eared Duck, another opportunistic breeder of inland Australia. The testes contain nodules of large seminiferous tubules filled with spermatids or spermatozoa at all times of the year. The adjacent seminiferous tubules contain resting spermatagonia only. The nodules vary in size depending on environmental conditions. A further adaptation to opportunistic breeding is a very short generation time, making it possible (in a good year) for the young to reproduce in the year in which they were hatched. Thus, Zebra Finches are able to breed within 90 days of hatch (Sossinka, 1980).

Rainfall, temperature and food supply are the main proximate inductive factors used to initiate breeding in opportunistic breeders. The relative importance of each depends on the species. The importance of rainfall as a proximate inductive factor in the initiation of breeding in Australian species has been over-emphasized (Davies, 1979). However, rainfall certainly plays a role. For example in Grey Teal, another Australian opportunistic breeder, held in captivity in natural lighting, both testicular weight and courtship display behaviour increase dramatically after a period of heavy thunderstorms (Figure 7.1).

Observations on opportunistic breeding in the Red-billed Quelea illustrate the importance of the maternal food supply as a proximate inductive factor in initiating breeding (Jones and Ward, 1976). In East Africa, this species is an 'itinerant breeder' which breeds during the rainy season. The birds migrate, following the rains as they pass across the

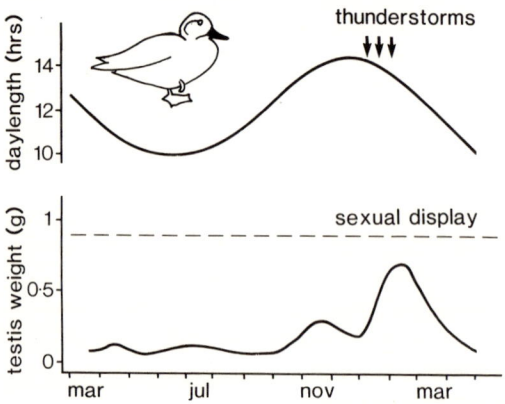

Figure 7.1 The stimulatory effect of thunderstorms on sexual display and testicular weight in captive Grey Teal exposed to natural changes in daylength in Canberra, Australia. The horizontal dotted line shows the testis weight at the production of the first spermatozoa. (From Braithwaite, 1976.)

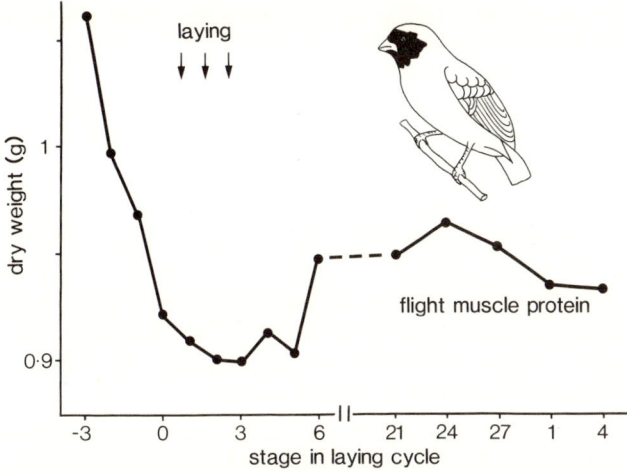

Figure 7.2 Flight muscle protein reserves in female Red-billed Quelea during the period before and after laying a clutch of eggs. The accumulation of these reserves is essential before breeding can be attempted. (From Jones and Ward, 1976.)

continent, stopping to breed when the opportunity arises. Each breeding attempt depends principally on an approximately 80% increase in the labile component of the muscle protein in the female. This is required for egg formation (Figure 7.2). Less thorough observations on Australian opportunistic breeders such as the Budgerigar (Wyndham, 1980) and Zebra Finch (Davies, 1979) also suggest that the timing of the breeding season is more directly related to the food available to the parents than to rainfall or temperature.

7.1.4 Tropical seasonal breeders

Although breeding is non-seasonal in some tropical birds, in many it is seasonal, depending primarily on the pattern of rainfall. In areas where there are pronounced wet and dry seasons, breeding tends to be associated with the rainy season, while in lowland rain forests, birds tend to breed during the driest period of the year. For example, in the tropical rain forests of Sarawak breeding occurs after the north-east monsoons which occur between December and February. Breeding is stimulated by the increase in insects which occurs as the results of a new growth of vegetation, stimulated by the monsoons. The resulting breeding season for passerine birds is hardly any less sharply defined than in Britain (Fogden, 1972). As in the Red-billed Quelea, the proximate inductive factor most important in the initiation of seasonal breeding in these tropical species is the food supply. This allows the build-up of reserves of muscle protein required for egg formation (Ward, 1969).

7.1.5 Spring and summer breeding

As we move out of the tropics to higher latitudes, seasonal changes in the food supply become less dependent on seasonal rainfall and more dependent on temperature. The breeding season coincides with those months when the temperature is sufficiently high to stimulate the growth of vegetation and consequently increase the vegetable, insect and other food resources. This dependency of food supply on temperature results in marked effects of latitude, longitude and altitude on the timing of breeding. For example, spring arrives later and winter earlier with increasing latitude, and the pattern of season breeding is timed accordingly (Murton and Westwood, 1977; Perrins and Birkhead, 1983). Species feeding young on ground invertebrates or fresh sprouting vegetation tend to breed before species taking flying insects, which, in turn, tend to breed before species taking seeds and fruits. Non-specialist feeders can have longer breeding seasons than specialist feeders. Seasonal changes in temperature are directly related to daylength, and hence the timing of the breeding season is correlated with daylength. Spring and summer breeders use daylength as a proximate inductive factor to time the onset of the breeding period (see section 7.3).

(a) *Symmetrical breeding cycles.* A small number of species, all of them pigeons or doves, reproduce as long as the daylength remains above a critical threshold (Figure 7.3a). In the case of the Wood-pigeon this is 12–13 h and in the Stock Dove, 10–11 h. Consequently, the breeding season of the Stock Dove is longer than that of the Wood-pigeon.

(b) *Asymmetrical breeding cycles.* In most spring and summer breeders, the breeding season has an asymmetrical relationship with the annual cycle in daylength (Figure 7.3 b–d). The breeding season ends when daylengths are still longer than those which stimulated breeding. Seasonal breeding is thus terminated by the development of insensitivity to the stimulatory effects of long days or 'photorefractoriness'. The ecological significance of photorefractoriness can be readily appreciated: it allows breeding to occur early in

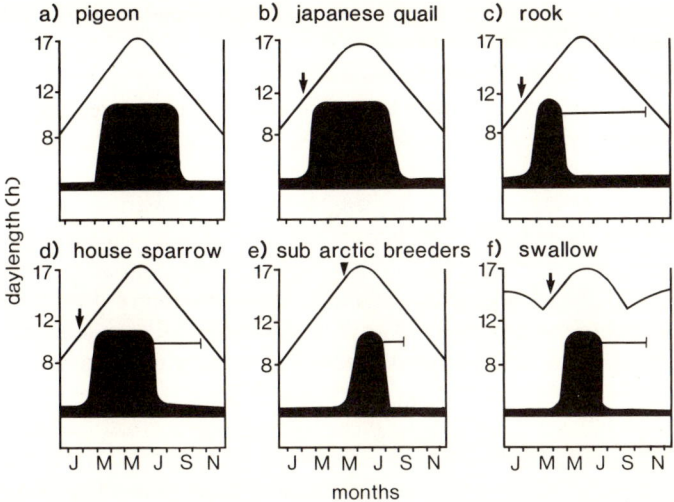

Figure 7.3 Patterns of seasonal breeding in birds reproducing at temperate and high latitudes. The shaded areas indicate the development of the reproductive system. The horizontal lines show the duration of absolute photorefractoriness. Arrows indicate the threshold of response to daylength. (Based on Lofts and Murton, 1968.)

the year to take advantage of specific food resources (e.g. Blue Tit: caterpillar; Mallard: young grass) and yet ends reproduction as soon as the food supply becomes scarce. It is also important in some migratory species, terminating breeding so that the young can grow sufficiently large to migrate to their wintering quarters.

The least dramatic form of photorefractoriness occurs in birds like Japanese Quail (Robinson and Follett, 1982) and domestic chicken (Sharp 1984) which start breeding in early spring and finish breeding in late summer or early autumn (Figure 7.3b). In the quail concentrations of plasma LH and FSH begin to increase in spring when the daylength exceeds 11.9 h. This results in the growth of the gonads, increased secretion of gonadal steroids and the development of steroid-dependent secondary sexual characters and behaviour. In late summer the reverse changes occur once the daylengths fall below 14.7 h. The form of photorefractoriness of the type found in quail is termed 'relative photorefractoriness'. This is because in late summer the birds become only relatively (compared with the early spring) insensitive to daylength. If the daylength is artificially kept longer than 14.7 h, photorefractoriness will not develop and the birds will remain in breeding condition indefinitely.

The most extreme forms of photorefractoriness are found in birds like the Rook and Mallard which begin to breed in the early spring and finish breeding in early summer (Figure 7.3c). For example, rooks breeding in north-east Scotland lay their eggs in March and April and gonadal development must therefore begin in February. The gonads begin their development at this time in response to an increased secretion of LH and FSH which occurs after the daylengths have increased to about 10.5 h (Figure 7.4). In May, the birds become photorefractory, and the concentrations of plasma gonadotrophins fall resulting in gonadal regression (Figure 7.4). Photorefractoriness is maintained for more than 4 months, until September, when there is an autumnal increase in the secretion of gonadotrophins, resulting in autumn sexuality (section 7.1.6). Rooks show

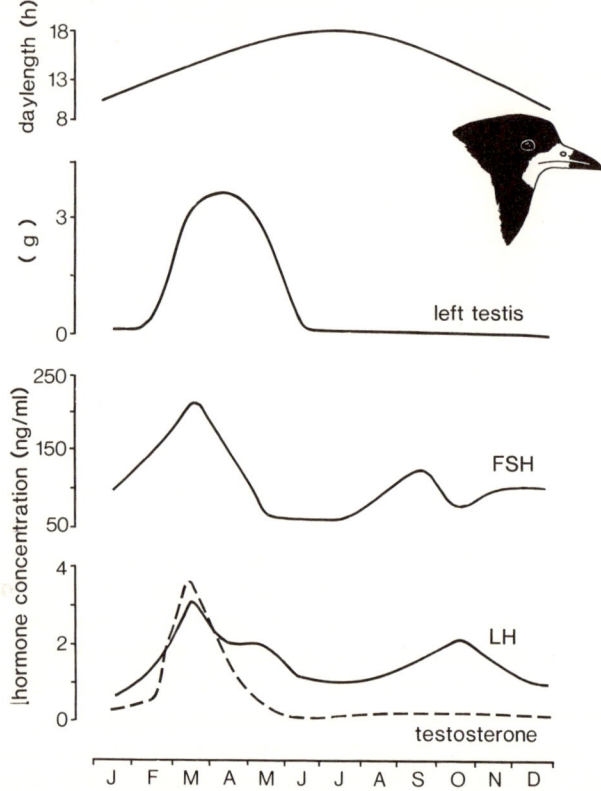

Figure 7.4 Seasonal changes in testis weight and concentrations of plasma hormones in the Rook. (From Lincoln et al., 1980.)

a form of photorefractoriness known as 'absolute photorefractoriness' because gonadal activity cannot be reactivated during the refractory period by further increasing the daylength, or by exposure to continuous light.

Some birds can produce several broods in a season and so may start breeding early and delay the development of absolute photorefractoriness until late summer or early autumn (Figure 7.3 d). Birds falling into this category include the House-sparrow, Goldfinch and Starling (Lofts and Murton, 1968).

At very high latitudes, spring comes late and birds must delay the beginning of the breeding season. This may be achieved by increasing the threshold of response to the daylength required to stimulate the growth of the gonads. The breeding period may thus become almost symmetrical around the summer solstice. Examples of this type of breeding cycle are to be found in single-brooded sub-arctic species including the Junco, the northern forms of White-crowned Sparrows and Willow Ptarmigan. In the high Arctic, for example on Spitzbergen (80°N), daylengths become continuous well before the arrival of the arctic spring. Here, proximate factors such as the maternal food supply and the presence of a mate may play a more important role in the initiation of breeding in the resident species (e.g. Spitzbergen Ptarmigan) than is the case in species breeding at lower latitudes (section 7.2). However, birds like the Spitzbergen Ptarmigan, are strongly photoperiodic and terminate breeding by the development of absolute photorefractoriness.

Transequatorial migrants such as the Bobolink and swallow also have breeding seasons which are almost symmetrical around the summer solstice, but they are exposed to a much more complicated pattern of seasonal changes in daylength than are other birds we have mentioned (Figure 7.3 f). They may experience continuous long photoperiods and some have evolved breeding cycles which have a high photoperiodic threshold requirement for the initiation of breeding. This high threshold serves two purposes: (1) it ensures that the gonads develop at the appropriate time on the breeding grounds; and (2) it prevents gonadal development during the period that the birds experience increasing daylengths in their wintering grounds. As in other migratory species, photorefractoriness must develop in time for the young to be sufficiently large to migrate to the wintering grounds. But unlike many birds with a photorefractory phase, photorefractoriness must be terminated in these transequatorial migrants by exposure to relatively long daylengths (i.e. more than 12 hrs). The problem faced by migrating species in the timing of seasonal breeding may not always be resolved, or even resolvable, by alterations in photoperiodic threshold. As we shall discuss later (section 7.4), they may adopt a strategy involving autonomous rhythms of reproductive activity.

7.1.6 *Autumn and winter breeders*

Although autumn and winter breeding is not common in birds, it does occur in a wide variety of circumstances. In the British Isles, Morley (1943) lists 68 species which show a resurgence of sexual activity during the autumn. Occasionally this may lead to egg-laying and production of young. This trend towards autumn breeding becomes more marked as we move towards the tropics, where the milder autumns allow some birds to reproduce for a second time in one year. This bimodal breeding pattern may be possible because the shortening days of autumn are able to terminate photorefractoriness while at the same time are still long enough to stimulate the secretion of gonadotrophins. This explanation cannot hold, however, for species which do not have periods of absolute photorefractoriness (Figure 7.3a, b). In these species it is likely that the basically unimodal pattern of seasonal breeding is converted to a bimodal pattern of seasonal breeding by the removal of proximal inductive factors operating during midsummer. Such factors may include shortage of food which override the stimulatory effects of daylength.

A few species breed exclusively in the autumn or winter. Amongst these are certain birds of prey (e.g. Eleonora's Falcon from the Mediterranean) whose breeding season is timed to coincide with the migration of its prey. The timing of the breeding season in the Kestrel can be advanced by as much as one month by the provision of additional food (Drent and Daan, 1980). It is therefore likely that, in autumn-breeding birds of prey, the food supply is a major proximate factor stimulating reproduction. Winter breeding is found in the northern hemisphere in the Crossbill whose chief food resource is pine seeds. Breeding in this species is induced by the good crop of pine cones, and is essentially opportunistic. In the southern hemisphere, breeding starts in some colonies of Silver Gulls nesting in Western Australia in the autumn and continues through the following winter and spring (Wooller and Dunlop, 1979). The same species is a spring breeder in some other parts of Australia. The Antarctic provides a further example of a short-day avian breeder, the Emperor Penguin, which lays its eggs in midwinter (Le Maho, 1977).

Autumn breeding without any associated spring breeding is also found in the sub-tropics amongst the munias (Estrilidean finches). When breeding in northern India these birds depend exclusively on grass seeds and crop grains for food (Chandola *et al.*, 1983). Since this food resource depends on the monsoons which occur in late summer, breeding must occur in the autumn. The physiological factors controlling seasonality in autumn breeders like Silver Gulls, Emperor Penguins and munias probably involve the use of environmental factors which phase autonomous rhythms of reproductive activity (section 7.4).

7.2 Proximate factors initiating breeding

Proximate inductive environmental factors are classified in relation to their functions. These provide (1) initial predictive information; (2) essential supplementary information; and (3) synchronizing and integrating information (Wingfield, 1983).

7.2.1 *Initial predictive information*

Initial predictive information is used by birds to bring them into the physiological state in which nesting may begin. Outside the tropics, seasonal changes in daylength provide a particularly useful source of initial predictive information but, with the exception of most game-birds, an increase in daylength is not sufficient in itself to induce full breeding condition. For example, the White-crowned Sparrow, a migratory species, does not breed in captivity even when exposed to very long daylengths (e.g. 20 h light per day). Although spermatogenesis is completed in photo-stimulated captive male White-crowned Sparrows, the testes are smaller,

Table 7.1 Testis size and concentrations of LH and testosterone in captive and free-living White-crowned Sparrows (from Wingfield, 1983).

	20L:4D[1]	Outdoor aviaries[2]	Arrival on territory[3]	Courtship and nesting[3]
LH (ng ml^{-1})	0.85	3.80	4.74	5.03
Testosterone (ng ml^{-1})	0.80	0.99	2.87	4.19
Testis weight (mg)	260 (max)	364 (max)	339	442 (max)

[1] Birds exposed to a constant 20 hour day.
[2] Birds exposed to natural lighting.
[3] Birds captured and sampled in the field.

and concentrations of plasma testosterone and LH are lower, than in birds captured in the field during the breeding season (Table 7.1). Similar differences in LH and gonadal hormone concentrations are also seen in captive and free-living female White-crowned Sparrows where the final stages of yellow-yolky ovarian follicular growth do not occur in the captive birds (Table 7.2). Thus, stimuli in addition to daylength are necessary for the final development of the gonads.

Changes in daylength provide a uniquely valuable form of initial predictive information. Unlike most other sources of initial predictive information, seasonal changes in daylength do not vary from year to year and can, therefore, be used to predict reliably when the food supply will increase. Photoperiodic species use this source of information to ensure that their young are produced to coincide with the increase in food supply. Early breeding is of considerable adaptive significance: it increases the chances of rearing two or more broods or renesting if a clutch is lost. Further, early-hatched chicks have a better chance of survival than chicks hatched later in the season (Perrins and Birkhead, 1983). In order to have young in the nest when the food supply is increasing, the reproductive system must begin to develop many weeks in advance. For example, in the Willow Ptarmigan living in northern Norway, concentrations of plasma gonadotrophins begin to increase in late March to allow time for the spermatozoa and yellow-yolky follicles to grow (Stokkan and Sharp, 1980). Egg-laying begins in May and a further period is required to complete the clutch and incubate it before producing chicks in June to coincide with the sprouting of the vegetation and increased abundance of insects. Thus, to hatch chicks to coincide with the increase in food supply, the prerequisite endocrine changes must be initiated more than 8 weeks previously. This

Table 7.2 Ovarian development and concentrations of LH and oestradiol in captive and free-living White-crowned Sparrows (from Wingfield, 1983).

	20L:4D[1]	Outdoor aviaries[2]	Arrival territory[3]	Courtship and nesting[3]
LH (ng ml^{-1})	1.99	1.80	3.00	40–10
Oestradiol (pg ml^{-1})	40	—	256	404
Ovary weight (mg)	50	30	40	500–1000
Diameter of largest follicle (mm)	3	3	3	10–12

[1] Birds exposed to a constant 20-hour day.
[2] Birds exposed to natural lighting.
[3] Birds captured and sampled in the field.

period, of course, differs between species, depending on the rate at which the reproductive system develops, clutch size and the incubation period.

Other types of initial predictive information such as rainfall, temperature, social interactions and maternal food supply are used to differing degrees by birds which can not rely on daylength serve in this role (sections 7.1.3, 7.1.4). In contrast with daylength, these sources of initial predictive information provide limited possibilities for birds to anticipate the breeding season. This problem can be partly solved by maintaining the reproductive system in a partially developed state, which may be achieved through an autonomous rhythm of reproductive activity (section 7.4).

In addition to providing initial predictive information to non-photoperiodic species, rainfall, temperature, social interaction and the maternal food supply may also provide all species with further classes of proximate inductive information required for the final reproductive effort.

7.2.2 Supplementary information

Essential supplementary information is required in addition to initial predictive information (section 7.2.1) to initiate the final growth of the gonads, and in particular, the formation of yellow-yolky follicles. In male birds which establish breeding territories, the successful acquisition of a territory and its defence results in an increased secretion of plasma LH and testosterone (Wingfield, 1983). For example, concentrations of these hormones increase within a few minutes of territorial disputes between male Red-winged Blackbirds (Harding and Follett, 1979). Paired females of many species also play a partial role in the defence of territory and this may be associated with increased concentrations of plasma testosterone, as in the male. Increased testosterone levels have been observed during territory formation in several species including the White-crowned Sparrow, the Pied Flycatcher, Mallard and Common Murre (Wingfield, 1983).

Since the female needs extra, high-quality food to produce her eggs, the maternal food supply can be a source of essential supplementary information. It is not merely a question of *quantity* of food required but its *quality* too. Thus, during egg formation, many herbivorous or granivorous species selectively eat sprouting shoots, leaves, flower buds or even algae in the spring. This new growth contains higher levels of proteins than old growth. Further, some supposedly herbivorous or granivorous species (e.g. Red Grouse, Song Sparrow, Red-winged Blackbird) may consume insects or other invertebrates, either as supplements to their diet, or even

Table 7.3 Effects of maternal diet on eggs laid by captive Red Grouse exposed to natural lighting in Scotland (57°N). (From Sharp and Moss, 1981.)

	Good diet[1]	Poor diet[1]
Number of hens	13	17
Mean date of first egg	2 May	2 May[2]
Rate of lay	2 days between eggs	3 days between eggs[3]
Mean egg weight (g)	21.9	21.2[3]
Percentage hatching	84	62[3]

[1]Good diet: 17.2% protein; poor diet 9.2% protein.
[2]Not different from birds fed a good diet.
[3]Significantly different from birds fed a good diet.

Table 7.4 Effect of chronic food restriction on mean concentrations of luteinizing hormone (LH) and testosterone (T) during photostimulation in White-crowned Sparrows (from Wingfield, 1980).

| | days of photostimulation | | | |
	0		5	
	LH (ng ml^{-1})	T (ng ml^{-1})	LH (ng ml^{-1})	T (ng ml^{-1})
Food				
ad libitum	0.4	< 0.1	1.5	1.15
5 g day^{-1}	0.3	< 0.4	2.0	< 0.04
2.5 g day^{-1}	0.2	< 0.1	2.5	< 0.04

exclusively, during the period of yolk formation. Experimentally, it has been shown in several species that supplementary feeding increases ovarian growth, and in some this may result in an advance in the date at which the first egg is laid (Perrins and Birkhead, 1983). However, this is not invariably the case. For example, the number and quality of eggs laid by Red Grouse is reduced by feeding a poor diet, but the date of the onset of lay is not affected (Table 7.3).

Although gonadal activity cannot be supported when food is severely limited, short periods of food restriction do not necessarily block photo-induced LH secretion (Table 7.4).

7.2.3 Synchronizing and integrating information

Once the nesting phase is initiated, the male must be ready to fertilize the female before the first egg is laid. This 'fine tuning' of male and female reproductive activity often depends on the presence of a completed nest and

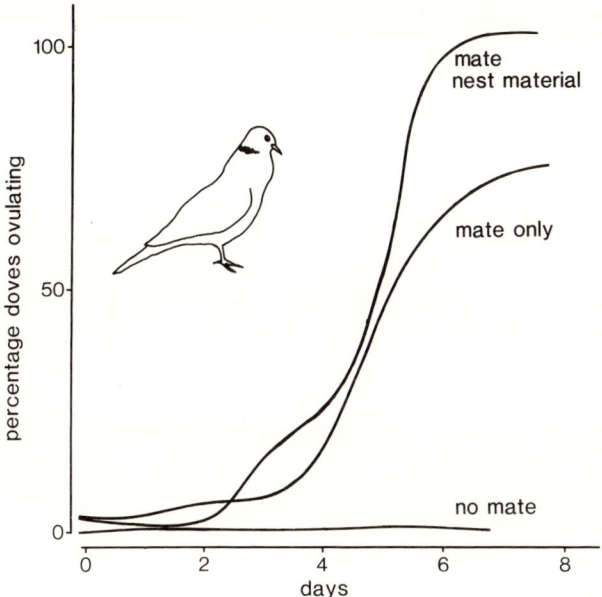

Figure 7.5 The induction of ovulation in isolated female Ring Doves by the introduction of a mate or mate with nest material. (From Lehrman *et al.*, 1961.)

usually involves some form of male/female interaction which stimulates further gonadotrophin release. These interactions have been carefully studied in several species notably the Ring Dove (Lehrman, 1965), Canary and Budgerigar (Hinde and Steel, 1978). An example, of the stimulatory effects of a mate and nest material on ovulation in the Ring Dove is shown in Figure 7.5.

The sexually active female also has a stimulatory effect on the endocrine status of the male (Wingfield, 1983). For example, concentrations of plasma LH and testosterone increase in male Ring Doves and pigeons exposed to females, while in male White-crowned Sparrows, a further increase in concentrations of plasma testosterone is observed after pair formation.

7.3 The mechanism of photoperiodic induction

The discovery that increasing daylength directly stimulates gonadal growth stems from Rowan's investigations (1925) into the control of migratory behaviour in the Junco (section 7.1.1). The mechanism involved has been most extensively studied in the Japanese Quail and White-crowned Sparrow and involves the bird's circadian system (Farner, 1964; Farner and Lewis, 1971; Follett, 1973, 1978, 1984; Follett *et al.*, 1977, 1981). An understanding of the nature of circadian rhythms and of their entrainment by '*Zeitgebers*' (Saunders, 1977) is therefore essential in any study of photoinduced gonadotrophin release. It could be argued that the effects of increasing daylength on gonadotrophin release are indirect and depend directly on increased food consumption and the consequent accumulation of energy reserves in the muscles. While increased food supply is undoubtedly an important source of predictive information for many birds (section 7.2), we have already noted in the White-crowned Sparrow that a decreased food supply does not necessarily prevent photo-induced gonadotrophin release (Table 7.4). Furthermore, in quail, gonadotrophin secretion increases steeply within the first day of exposure to long days (Figure 7.6). It is difficult to see how this increase could be mediated by a change in food intake.

We have seen how the secretion of gonadal steroids and gonadotrophins are maintained in a state of dynamic equilibrium (section 6.2.3). It is therefore possible that gonadotrophin secretion may be stimulated by increasing daylength as a consequence of a decrease in the sensitivity of the inhibitory feedback system. If this hypothesis is correct, then in the absence of the inhibitory feedback signal (i.e. gonadal steroids), gonadotrophin secretion should not be stimulated by increasing daylengths. Gonadotrophin secretion should not, therefore, be influenced by changes in daylength in castrated birds. This hypothesis has been tested experimen-

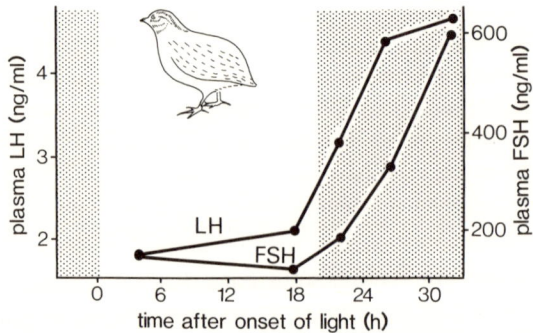

Figure 7.6 The effect of exposure to one long day (20 hours of light) on the secretion of FSH and LH in Japanese Quail. The birds had previously been exposed to 8 hours of light per day. (From Follett *et al.*, 1977.)

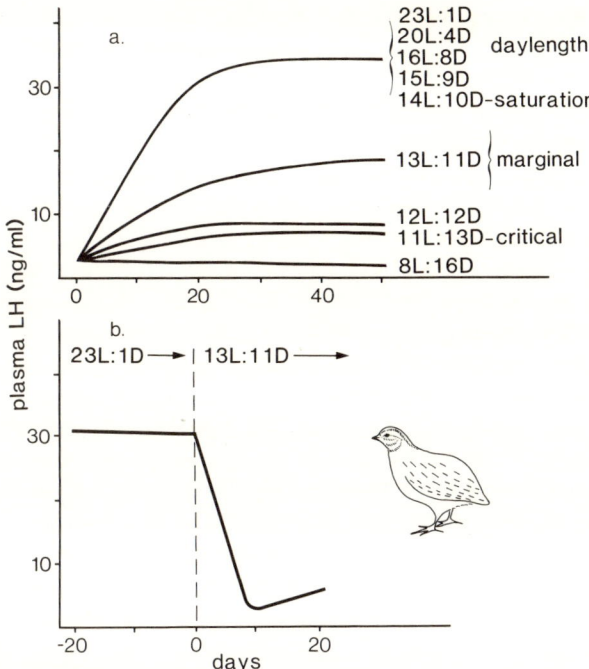

Figure 7.7 Changes in the concentrations of plasma LH in castrated Japanese Quail exposed for several weeks to (*a*) short days (8 h light :16 h darkness, 8L :16D) or to (*b*) long days (23 h light :1 h darkness) before transfer to the daylength shown. Note that the response of birds to 13L :11D depends on their photoperiodic history. A 13-hours day is stimulatory in birds previously exposed to 8L :16D but inhibitory in birds previously exposed to 23L :1D. (From Urbanski and Follett, 1983.)

tally and shown to be wrong (Figure 7.7). In castrated quail, gonadotrophin secretion is stimulated by increasing daylengths and inhibited by decreasing daylengths. Thus, changes in daylength appear to regulate the release of gonadotrophins by acting directly on the brain to increase or decrease 'hypothalamic drive'.

7.3.1 *The critical daylength*

The minimum daylength required to stimulate gonadotrophin secretion is termed the *critical daylength* and can be determined experimentally by transferring photosensitive birds from short days to a range of stimulatory daylengths. For example, when castrated quail are transferred from short days to long days, the minimum daylength required to stimulate LH secretion is between 11 and 12 h (Figure 7.7). Over a range of daylengths above the critical daylength, termed *marginal daylengths*, there is a direct relationship between daylength and concentrations of plasma LH. If the daylength is increased further a *saturation daylength* is reached which stimulates the maximum release of LH. The photoperiodic response is now 'saturated' and further increases in daylength have no additional stimulatory effects on LH secretion.

Species differences in critical, and marginally stimulatory, daylengths play an important role in the timing of the onset of seasonal breeding. There tends to be a direct relationship between critical daylength and the latitude at which breeding occurs. This relationship is illustrated by a study of the breeding seasons of several species of swans originating from different latitudes, and held in captivity in the United Kingdom (Figure 7.8). The Black Swan which normally breeds at 28°S has a short critical daylength, since it starts to breed in late January when the daylength is only about 10 h. At the other extreme, the Bewick Swan which normally breeds at 65°N has a long critical daylength, since it only begins to lay eggs after the daylength exceeds 16 h. The Mute and Coscoroba Swans, coming from intermediate latitudes, fall between these two extremes. One of the costs of the evolution of critical daylengths can be to limit the breeding range. Thus, birds like the Bewick Swan cannot breed at latitudes where the maximum seasonal daylength does not exceed 16 h.

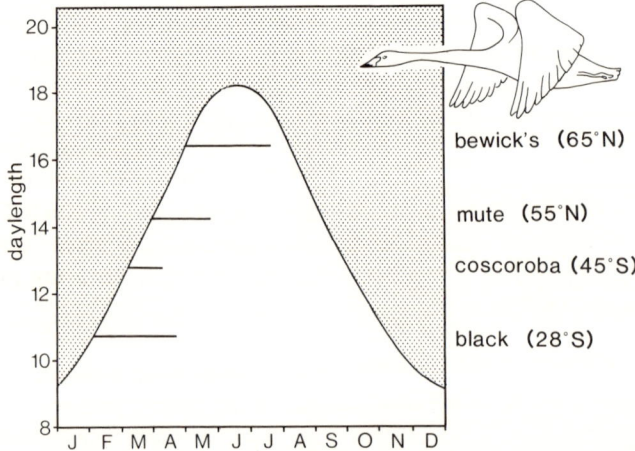

Figure 7.8 The relationship between the timing of seasonal breeding (horizontal lines) and the latitude of origin of four species of captive swans exposed to natural lighting at 51°N (UK). (From Murton and Westwood, 1977.)

Species differences in the range of marginally stimulatory daylengths may also be of adaptive significance. By increasing the range, the time taken to reach sexual maturity after the critical daylength is exceeded can be varied. For example, in quail exposed to natural lighting, this period is about 35 d, whereas in the White-crowned Sparrow it is about 90 d (Follett, 1984).

The critical daylength not only differs between species, but also changes in some species during the course of the breeding season. Let us consider the photoperiodic response of castrated quail again (Figure 7.7). As we have already seen, when the birds are transferred from short days (i.e. 8 h light per day) to long days, the critical daylength is 11–12 h and a 13 h day is regarded as a marginal stimulatory daylength. If the castrated quail are maintained for more than 2 months on a long day-length (23 h light per day) and then transferred to a 13 h day, concentrations of plasma LH drop to values well below those in birds transferred from an 8 to a 13 h daylength. Thus, in birds exposed to long days the critical daylength is longer than in birds exposed to short days. This observation provides an explanation for the development of relative photorefractoriness (section 7.1.5). In the spring the critical daylength is short and breeding is stimulated after it is exceeded. But continuous exposure to the long days of summer causes the critical daylength to lengthen. Thus in the autumn as daylengths decrease, the critical daylength is soon reached and as a result, breeding is terminated. It follows that exposure to short days during the late autumn and winter causes the critical daylength to shorten in preparation for the following spring.

The neuroendocrine mechanisms which determine the critical daylength are not fully understood. For example, in quail after transfer from a short to a long day, concentrations of plasma gonadotrophins do not increase immediately after the critical daylength (11.9 h) is exceeded (Figure 7.6). They begin to increase about 18 h after the beginning of the light period. This lag in photoperiodic response might reflect the way in which the bird's time-measuring system is organized, or sluggishness in the neuroendocrine pathways involved in the release of gonadotrophin releasing hormone (Follett *et al.*, 1977).

7.3.2 *The role of circadian rhythms*

The circadian rhythms involved in the regulation of photoperiodically-induced gonadotrophin release are a function of the central nervous system. Two important properties of these rhythms are:

160 AVIAN BIOLOGY

(1) They persist in the absence of *Zeitgebers* (e.g. in constant darkness).
(2) The phase relationship between a circadian rhythm and its entraining light-dark cycle is a function of the light: dark ratio.

Two basic circadian models, which are not mutually exclusive, are used to explain the mechanism of photoinduced gonadotrophin release (Figure 7.9).

(a) *The external concidence model.* In this model, gonadotrophin release occurs when light is coincident with a certain phase of a circadian rhythm of photosensitivity. This phase is termed *the photo-inducible phase* and, in the quail at least, is known to last between 3 and 6 hrs. The exact position of the photoinducible phase depends on the entraining properties of the light–

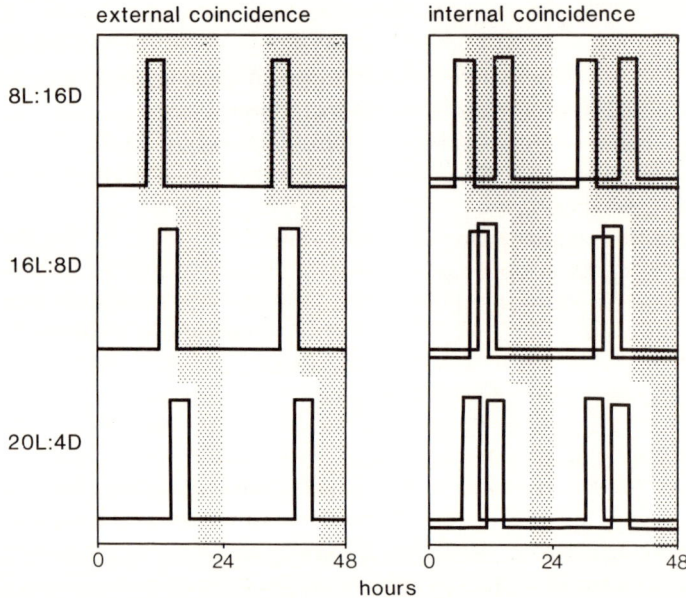

Figure 7.9 Diagram to illustrate the way circadian rhythms may be involved in the regulation of gonadotrophin secretion in birds. Two basic models are shown. In the external coincidence model a single circadian rhythm of photoinducibility is shown with a square waveform. The crest of the wave is the photoinducible phase: when this phase is coincident with light, gonadotrophin secretion is stimulated. In the internal coincidence model two circadian rhythms, again shown in a square waveform, are differentially entrained by a given light–dark cycle. As the light:dark ratio changes, the relationship between the two rhythms also changes. When the daylengths are stimulatory, the two rhythms are entrained into an inductive relationship which in this model is assumed to occur when the crests of the two rhythms coincide. (From Sharp, 1984.)

dark cycle (see (2) above) and is not fixed relative to dawn or dusk. However, as daylength increases the photoinducible phase moves out of the dark period into the light. Coincidence between light and the photoinducible phase induces the release of gonadotrophins. Thus, in the 'external coincidence model' light has two functions: it entrains the circadian rhythm of photosensitivity and induces the release of gonadotrophins.

(b) The internal coincidence model. This model is based on two assumptions: firstly, that photoinduced gonadotrophin release depends only on the entraining properties of daylength and does not require coincidence between light and a photoinducible phase; secondly, that the circadian rhythm of photosensitivity is the product of two or more circadian rhythms coupled with each other and to a master clock. The two circadian rhythms are entrained differently by the light–dark cycle, so that the phase relationship between them is constantly altering as the daylength changes. Photoinduced gonadotrophin release occurs when the two rhythms have a particular phase relationship—in this case when the two phase points overlap (Figure 7.9). In the model shown, the 20 h-day entrains the two circadian rhythms into a non-inductive relationship, thereby rendering the bird photorefractory.

On considering the photoperiodic responses of birds, the internal coincidence model offers more flexibility than the external coincidence model. In particular, it goes some way towards explaining the development of photorefractoriness. However, neither model is entirely satisfactory and they are only of value inasmuch as they suggest further experiments on the nature of the photoperiodic response.

Based on a theoretical knowledge of circadian rhythms, several types of experiments can be done to demonstrate that circadian rhythms are involved in the mechanism of photoinduced gonadotrophin release (Saunders, 1977; Follett, 1984). One of the most widely used is the so-called 'resonance' experiment. In a typical experiment, photosensitive birds are transferred from short days to lighting schedules in which a short period of 6 or 8 hours is combined with dark periods of up to 96 hours e.g. 8 hours light: 4 hours dark (8L:4D), 8L:16D, 8L:28D, 8L:40D, 8L:52D, 8L:64D, 8L:76D and 8L:88D. Gonadotrophin secretion is stimulated in cycles of 12, 36, 60 and 84 h but not in those adding up to 24, 48, 72 and 96 h. Thus, every 24 hours there is a period when light will stimulate gonadotrophin release. In other words, there is a rhythm of photoperiodic sensitivity which resonates with a period of about 24 hours. By definition such a rhythm is circadian in nature. A classical example of this type of experiment, using

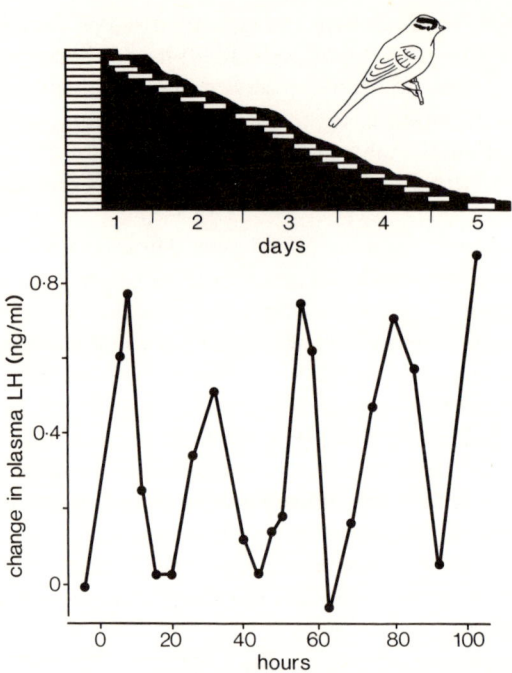

Figure 7.10 A 'resonance' experiment to show that birds make use of their circadian system to measure daylength. White-crowned Sparrows were kept on an 8 h day and a pre-experimental blood sample was taken early in the last light period before transfer to continuous darkness. At various times thereafter each bird was exposed to a single 8 h period of light. These lighting treatments are shown in the upper part of the figure. A second blood sample was taken a few hours after the experimental light pulse. The graph shows the change in plasma LH concentration between the two samples in relation to each treatment. The data shows a daily periodicity in photosensitivity to the 8 h pulse of light. It is not the amount of light which is important in triggering gonadotrophin release but when it falls relative to the underlying circadian rhythmicity within the bird. (From Follett *et al.*, 1974.)

White-crowned Sparrows, is shown in Figure 7.10. An extra refinement is added to the basic experimental model by measuring the photoperiodic response (an increase in LH secretion) after exposure to only one lighting cycle. In this way, problems of interpretation caused by the entraining effects of repeated lighting/cycles are avoided.

7.4 Autonomous rhythms and scotorefractoriness

As we have discussed (section 7.2), one of the keys to a successful breeding strategy is to be able to predict and prepare for the time when the food supply increases. Changes in daylength are a most valuable source of proximate information since they provide advanced warning of a forthcoming breeding season. We have discussed the way in which daylength provides *proximate inductive information* but it can also provide *proximate phasing information* for those species which process autonomous rhythms of reproductive activity. Such rhythms provide added insurance that breeding will be timed precisely, thus protecting the bird against inappropriate *proximate inductive information*. This is particularly important in transequatorial migratory species or in migratory birds which winter close to the equator.

There are also situations in which seasonal changes in daylength can not be used as a source of *proximate inductive information* to stimulate reproductive activity. The most obvious is in the tropics where seasonal changes in daylength do not occur, or are very small. Outside the tropics an inductive effect of daylength on reproduction cannot be used to time breeding in autumn or winter breeders or to stimulate territorial or other forms of pre-nuptial behaviour during the winter months in spring breeders. In these situations birds may use autonomous rhythms of

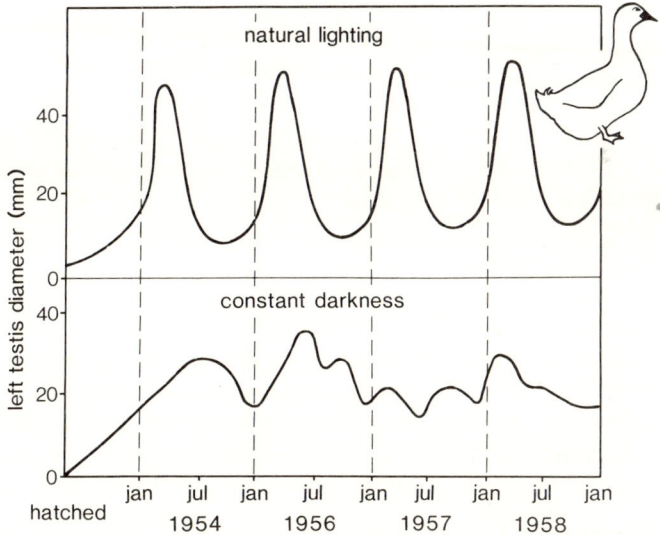

Figure 7.11 Autonomous rhythms of testicular size in Pekin drakes maintained in constant darkness. The testicular sizes of the control birds exposed to natural lighting are shown in the upper graph. (From Assenmacher, 1974.)

reproductive activity to ensure that breeding occurs at the most appropriate time.

The first demonstration of an autonomous rhythm of reproductive function came from a study of testicular size in Pekin ducks kept for four years in continuous darkness (Figure 7.11). The mechanism responsible for these rhythms is unknown and several hypotheses have been advanced (Gwinner, 1981).

1. *Frequency demultiplication of circadian rhythms.* Just as an electrical clock produces a 24 h rhythm from the 50 or 60 cycles s^{-1} frequency of the electrical current, an autonomous cycle might be generated from circadian frequencies.

2. *Autonomous rhythm of a circadian rhythm of photosensitivity.* This is based on the external coincidence model for photoinduced gonadal growth (section 7.3.2). It is suggested that the phase relationship between the circadian rhythm of photosensitivity and the light–dark cycle to which it is entrained is subject to autonomous variations.

3. *Autonomous variations in the internal circadian system.* This is based on the internal coincidence model for photo-induced gonadal growth (section 7.3.2). It is suggested that the phase relationship between two or more circadian rhythms involved in the control of gonadotrophin release exhibit autonomous variations.

4. *A sequence of stages.* It is suggested that autonomous rhythms are not 'true' rhythms but only a sequence of linked stages, each one taking a given amount of time to complete and then leading into the next, with the last stage linked back to the first again.

7.4.1 Tropical opportunistic breeders

We have given a descriptive account of autonomous breeding cycles in some tropical sea-birds and have cited evidence for such cycles in some opportunistic breeders (section 7.1.3). In one of these birds, the Red-billed Quelea, there is a precise 12-month cycle of testicular growth and regression with a constant 42-day refractory period (Lofts, 1964). In the sea-birds, the proximate phasing factor is that of social facilitation (section 7.1.3). The nature of the proximate phasing factors, which may phase autonomous rhythms of reproductive activity where they exist, is unknown in most opportunistic breeders. In view of the adaptive advantage of being able to predict and prepare for the breeding season, it would not be surprising if many tropical seasonal breeders make use of autonomous rhythms of reproductive activity. However, this possibility has not been fully investigated.

7.4.2 Migratory species

Autonomous rhythms of testis size, moult, body weight and pre-migratory restlessness occur in several species of migratory warblers maintained on constant 10-, 12- or 20-hour daylengths for two years or more (Berthold, 1975). These rhythms have a period of about 10 months and are referred to as 'circannual', being close to, but not equal to, an annual cycle of 12 months. The precision and persistency of these circannual rhythms in migratory warblers is related to the intensity of the migratory instinct. For example, circannual rhythms are easily demonstrable in the Garden Warbler, a long-distance migrant, while in the closely related Chiffchaff, a partial migrant, they are not. This observation emphasizes the importance of autonomous rhythms to long-distance migrants.

Seasonal changes in daylength probably provide proximate phasing information for these autonomous rhythms. For example, in the Starling, a partial migrant with an autonomous rhythm of reproductive activity, cycles of testicular growth and regression can be synchronized by cycles of increasing and decreasing daylength of between 3 and 12 months (Gwinner, 1981).

7.4.3 Autumn and winter breeders

The evidence for autonomous rhythms in autumn and winter breeders (section 7.1.6) is limited. However, the evidence available for the Spotted Munia is particularly persuasive (Figure 7.12). Under natural lighting the testes in this species begin to enlarge in July and August, and full testicular function is maintained during September and October. Experimentally, gonadal development is not directly responsive to a change in daylength. When these birds are maintained on a 12-hours daylength or in constant light for two years, two cycles of testicular growth and regression are observed with a period of about 10 months. These cycles are out of phase with the natural cycle, and hence it is evident that they are truly autonomous and are not driven directly by some unidentified proximate inducing factor.

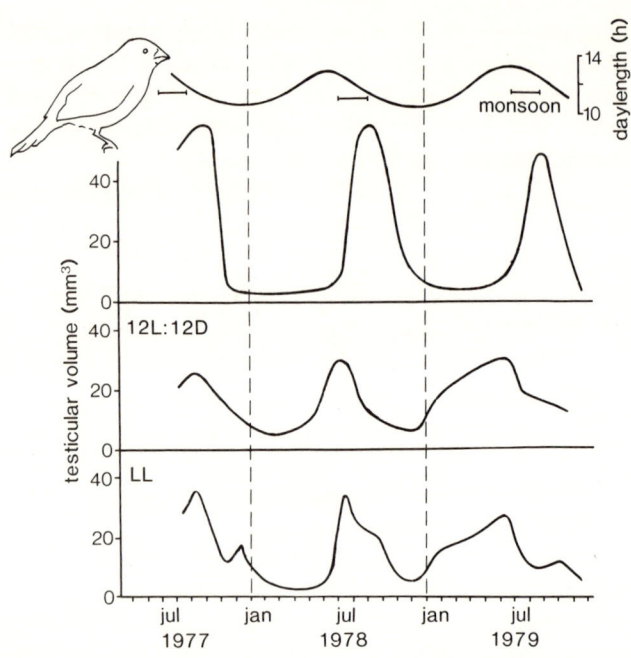

Figure 7.12 Autonomous rhythms of testicular size in Spotted Munia from northern India (25°N). The birds were kept on natural lighting (upper graph) or exposed to 12 h light per day (12L:12D) or to continuous light (LL) for more than 2 years. (From Chandola *et al.*, 1983.)

The proximate factors responsible for phasing the autonomous rhythm in the Spotted Munia are not known with certainty (Chandola *et al.*, 1983). One could be the annual change in daylength, since the autonomous growth of the testes is advanced if birds are transferred in March to an artificially reversed annual lighting cycle (i.e. transferred from increasing to decreasing daylengths). It is also possible that seasonal changes in food availability phase the autonomous rhythm: food intake drops to its lowest level just prior to the breeding season. The drop in food intake is correlated with a fall in the concentrations of plasma thyroxine which, as mentioned in section 6.3, could stimulate gonadal growth.

7.4.4 *Scotorefractoriness: its role in autumn and winter sexuality*

The term 'scotorefractoriness' is used to describe the condition in which long-day breeders become insensitive to the inhibitory effects of short days.

Many temperate-zone breeders, amongst others quail, ducks, Slate-coloured Juncos, Starlings and partridges, come into full breeding condition after prolonged exposure to short days (Sharp, 1984). The development of *partial* scotorefractoriness is an important feature of the annual cycle of many temperate-zone birds and results in an increase in the concentrations of plasma gonadotrophins in the autumn and winter. The increase in concentrations of plasma gonadotrophins in the autumn and winter is sufficient to stimulate steroidogenesis, but not (or only rarely) full gametogenesis. The increased plasma concentrations of gonadal steroids, in turn, induce territorial behaviour and pair-bond formation. As daylengths increase in spring, the critical daylength is finally reached, resulting in a further increase in gonadotrophin release and the completion of gametogenesis.

The development of scotorefractoriness is an autonomous process and the mechanisms responsible may have much in common with those involved in the generation of autonomous rhythms.

7.5 Absolute photorefractoriness

We have discussed the importance of the development of absolute photorefractoriness as a strategy to terminate breeding well before the food supply required for reproduction begins to fail (section 7.1.5). Now we turn our attention to the nature of the physiological mechanisms responsible for the development of the condition. Photorefractoriness has probably evolved independently in different groups of birds and hence the mechanism involved may differ between species (Farner *et al.*, 1983). In many species (e.g. White-throated Sparrow) the development of the condition is intimately associated with migratory behaviour.

7.5.1 *Control by the central nervous system*

The secretion of FSH and LH is inhibited during the development of photorefractoriness (e.g. Figure 7.4). Thus, photorefractoriness is not due primarily to a failure of the gonads to respond to gonadotrophins. In fact, gonadal growth can be induced in photorefractory birds by injections of gonadotrophins. A failure of the pituitary gland to respond to GnRH can also be discounted since gonadotrophin secretion is readily stimulated in photorefractory birds by injections of synthetic mammalian GnRH. We therefore conclude that photorefractoriness is caused by a decrease in the release of GnRH from the hypothalamus.

7.5.2 Dependence on daylength

The rate at which birds become photorefractory depends on the lighting pattern to which they are exposed. If birds are transferred from short days to a range of fixed daylengths above the critical daylength there is an inverse relationship between the period that the birds are in breeding condition and the daylength. For example, in the Starling, photorefractoriness (as indicated by testicular regression) begins 2 months after transfer from short days to a 15-hour daylength but is delayed for 3 months in birds transferred to a 13-hour daylength (Hamner, 1971). This relationship can have an adaptive significance for species like the House Sparrow, which breed over a wide range of temperature-zone latitudes. With decreasing latitude, the maximum daylength reached in the summer decreases while the seasonal increase in food supply increases. Thus, at lower latitudes, extended breeding seasons are possible. In the House Sparrow, the decrease in daylength with decreasing latitude results in a delay in the development of photorefractoriness and hence extends the breeding season (Table 7.5).

Exposure to stimulatory daylengths just above the critical daylength may even prevent the development of photorefractoriness. For example, photorefractoriness does not develop in Starlings and Canaries exposed to 11 hours of light per day or in Willow Ptarmigan exposed to a 14-hour day (Sharp, 1983).

Absolute photorefractoriness is terminated by exposure to short days or by reducing the light intensity. In birds such as the Willow Ptarmigan, White-crowned Sparrow, Starling and Canary, photorefractoriness can be terminated by exposure to a 6- or 8-hour days for 5–6 weeks. The minimum

Table 7.5 Effect of latitude on duration of breeding and absolute photorefractoriness in the House Sparrow (from Murton and Westwood, 1977).

Latitude (°N)	Spermatozoa production (days)	Photorefractory (days)
34	138	13
36	135	31
43	118	64
52	106	100

daylength required for the termination of photorefractoriness is longer than that required to stimulate or maintain gonadotrophin secretion in photosensitive birds. Thus, in the House Finch, photorefractoriness can be terminated by exposure to either a 6- or a 14-hour day (Hamner, 1968).

7.5.3 Role of the circadian system

We have discussed how birds use their circadian system to initiate breeding in response to increasing daylength (section 7.3.2). The problem now arises, how can long days switch from stimulating to inhibiting gonadotrophin release? If we consider the two models for photoinduced gonadotrophin release (Figure 7.9) the external coincidence model offers no solutions. It could be argued that photorefractoriness develops because the photoin-

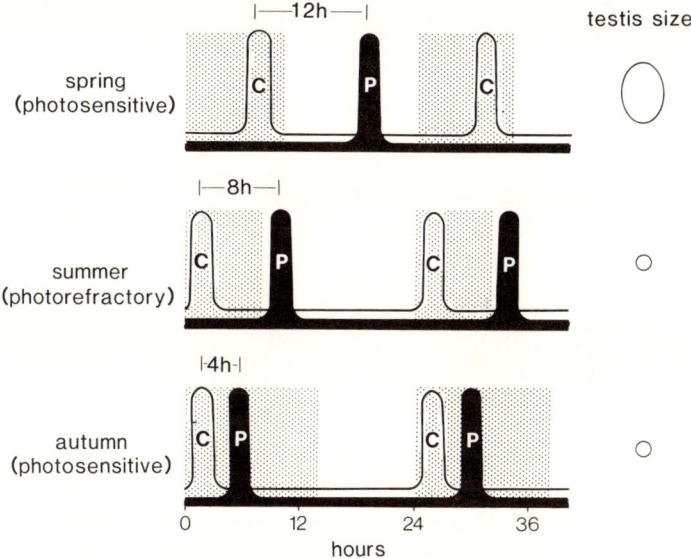

Figure 7.13 Meier's hypothesis explaining how seasonal changes in daily rhythms in the secretion of corticosterone (C) and prolactin (P) reflect the circadian rhythms controlling seasonal breeding, migration and pre-migratory fattening. The stippled areas represent the hours of darkness. In the spring, the daily rhythm in plasma corticosterone secretion peaks just before dawn whereas in the summer and autumn it peaks just after sunset. In the spring the daily rhythm in plasma prolactin peaks in the late afternoon, but during the summer and autumn it is progressively delayed until it occurs towards the beginning of the night. The time interval between the peaks of corticosterone and prolactin reflects the reproductive status of the bird. (Adapted from Meier and Ferrell, 1978.)

ducible phase moves into the hours of darkness. However, this seems unlikely, since in all absolutely photorefractory species studied, photorefractoriness persists when the birds are exposed to continuous light. Under such circumstances, in terms of the external coincidence model, the photoinducible phase should be coincident with light at all times.

The internal coincidence model offers a limited explanation for the development of photorefractoriness and for migratory behaviour (Meier and Ferrell, 1978). The two circadian rhythms involved in seasonal breeding may generate daily rhythms in the secretion of prolactin and corticosterone. As is suggested by the internal coincidence model, the phase relationships between these rhythms change seasonally (Figure 7.13). During the breeding season, peak concentrations of plasma corticosterone occur just before sunrise and peak concentrations of prolactin, some 12 h later, in the late afternoon. During the development of photorefractoriness the phases of both rhythms are advanced, with that of prolactin being more advanced than that of corticosterone. So, in the photorefractory bird, the peak concentrations of prolactin occur in the early morning, some 8 h after the corticosterone peak. In the autumn, when photorefractoriness is terminated, the two peaks come even closer together, being separated by only 4 h. Experimental support for this hypothesis comes from studies, chiefly on White-throated Sparrows, in which appropriately timed daily injections of corticosterone and prolactin can mimic the proposed seasonal rhythms in the secretion of these hormones. Thus, in birds maintained in continuous dim light to eliminate photoperiodic entrainment, injections of corticosterone and prolactin 12, 8 and 4 hours apart induced spring-, summer- or autumn-like reproductive and migratory responses, respectively. A 12-hour relationship between the hormone injections induces gonadal growth, body fat deposition and northward-orientated migratory restlessness. Injections given 8 hours apart strongly inhibited gonadal growth and have no effect on migratory behaviour, thus simulating the photorefractory period. Injections given 4 hours apart have no effect on gonadal weight but induce body fat deposition and southward-orientated nocturnal restlessness. These observations have not been successfully applied to many other avian species, and hence the significance of the hypothesis in a broader context is in doubt.

7.5.4 *Some endocrine explanations*

There is no doubt that the development of photorefractoriness is caused by a central nervous mechanism involving the circadian system. However, it

(a) Endocrine exhaustion. Photorefractoriness could be due to the exhaustion of some elements in the neuroendocrine system controlling the release of gonadotrophin releasing hormone (GnRH). If GnRH release could be inhibited during photostimulation, the hypothesis suggests that photorefractoriness will not develop. This idea has been investigated in Willow Ptarmigan, White-crowned Sparrows and Canaries (Farner et al., 1983). In males of these species, it is possible to suppress completely the stimulatory effects of increasing daylength on LH secretion, and hence the development of the testes, by the inhibitory feedback effects of subcutaneous implants of testosterone (see section 6.2.3). Removal of the implants at intervals is followed by increases in LH secretion, but removal after the control birds have become photorefractory is followed by no increase in LH secretion or in the growth of the testes. Thus, even though LH release and by implication, GnRH release was not stimulated by long days, the birds still become photorefractory. It thus seems that photorefractoriness is not due to a simple form of exhaustion of the neuroendocrine system.

(b) Changes in responsiveness to inhibitory steroid feedback. The fall in concentrations of plasma gonadotrophins during the development of photorefractoriness could be due to an increase in the responsiveness of the hypothalamus to inhibitory feedback steroids. In principle, this hypothesis is easy to test. If inhibitory feedback steroids are removed,

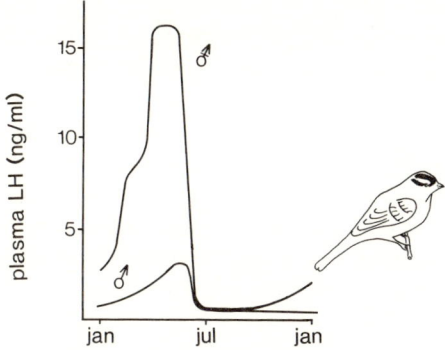

Figure 7.14 Seasonal changes in concentrations of plasma LH in intact and castrated White-crowned Sparrows exposed to natural lighting. The castrated birds become photorefractory, as indicated by the fall in LH levels which parallels that in the intact controls at the end of the breeding season in June. (From Mattocks et al., 1976.)

seasonal changes in gonadotrophins should follow the changes in daylength and not fall at the time intact control birds become photorefractory. In several passerine species including the Canary, Starling and White-crowned Sparrow (Figure 7.14) plasma LH levels *do* fall when the intact control become photorefractory (Farner *et al.*, 1983). Hence, in these birds, castration does not prevent the development of photorefractoriness. In non-passerine species including the Willow Ptarmigan (Stokkan and Sharp, 1984) and Mallard (Haase *et al.*, 1982), the situation is not so clearcut. Seasonal changes in concentrations of plasma LH in castrates do not closely follow the pattern seen in the intact controls. Indirect evidence of a role for gonadal steroids in the development of photorefractoriness is also found in the sub-tropical Weaver Finch (Pavgi and Chandola, 1981). In this bird, yellow, nuptial plumage develops when concentrations of plasma LH are high. In castrated Weaver Finches yellow plumage continues to be produced throughout the long days of summer and autumn, whereas in intact controls, brown plumage is produced after the development of photorefractoriness in early autumn. It thus seems that, in terms of LH secretion, the castrated bird does not become photorefractory.

(c) *Increased secretion of thyroid hormones or prolactin.* We have seen that thyroxine (section 6.3) and prolactin (section 6.6) exert an inhibitory effect

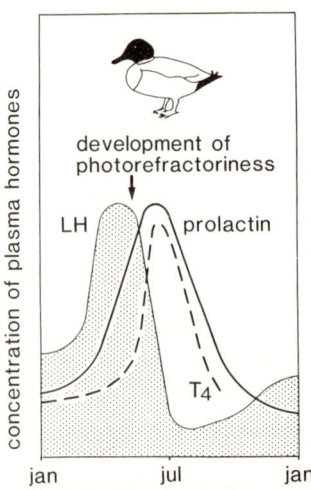

Figure 7.15 Seasonal changes in concentrations of plasma LH, prolactin and thyroxine (T_4) in the mallard. The development of photorefractoriness is associated with increases in concentrations of plasma prolactin and T_4. (P.J. Sharp, unpublished observations.)

on gonadotrophin secretion. It is therefore possible that an increase in the secretion of either or both these hormones could play a role in the initiation of photorefractoriness. In a wide range of species including the Starling, Capercaillie, Rook and Mallard, the concentrations of plasma prolactin begin to increase during the development of photorefractoriness (Figure 7.15). It remains to be determined whether this increase in prolactin secretion is a consequence of the development of photorefractoriness or a cause. One problem arises in multi-brooded species. If an increase in plasma prolactin *causes* photorefractoriness, why do these birds not become photorefractory during the time they are incubating their first clutch when prolactin secretion is high?

A role for thyroxine in the development of photorefractoriness is an attractive possibility in those species in which concentrations of plasma thyroxine increase at the end of the breeding season. These include the Starling and Mallard (Figure 7.15) but not the Willow Ptarmigan (Klandorf *et al.*, 1982) or Rook (Lincoln *et al.*, 1980). In the Starling, and several other species (section 6.3) the administration of thyroxine induces gonadal regression and importantly, it also induces photorefractoriness and an associated increase in the secretion of prolactin (Goldsmith and Nicholls, 1984). It thus seems that thyroxine may play an important role in the induction of photorefractoriness in some species, but not necessarily in others.

CHAPTER EIGHT

APPLIED ASPECTS

8.1 Domestication

Birds are bred in captivity for many reasons. They are kept as pets for their song or the appearance of their plumage and are reared for sport. But numerically most are reared for meat or for eggs. Domestication occurs by artificial selection for many generations and results in races which are entirely dependent on man for their survival. Amongst the species which have been domesticated are the Red Jungle Fowl, Quail, Turkey, Greylag Goose, Mallard, Rock Pigeon, Budgerigar, Canary, Bengalese Finch, Zebra Finch and Rice Finch. Economically the most important of these are the domestic chicken, turkey and duck. Their origins are reviewed by Herre and Rohrs (1983).

8.1.1 *Chickens*

Domestic chickens are descendants of jungle fowl which occur in India and South-east Asia. Today there are four races of jungle fowl, but in the past there may have been others, now extinct, which also formed part of the domesticated chicken's ancestry. These birds would have been characterized by their larger size and would have given rise to the heavy Asiatic breeds including the Cochins and Brahmas (Hutt, 1949). Birds descended from the lighter races of jungle fowl are termed Mediterranean breeds, the best known of which is the White Leghorn.

The jungle fowl is a sub-tropical and equatorial species. To the north of its range it is a long-day breeder, beginning to lay in February. In equatorial regions, it tends to be a dry-season breeder or to have no seasonal pattern of breeding (Nishida, 1980). A flock of wild jungle fowl usually consists of a male leader, one to seven females and, sometimes, a few younger males. A female weighs about 0.75 kg and lays a clutch of 5–7 eggs (Nishida, 1980). Most domestic breeds of chicken are 'native' varieties, selected as often for their fighting abilities as for their egg-laying capacity.

Table 8.1 A comparison of egg production, body weight and feed consumption in three 'unimproved' breeds of domestic chickens exposed to seasonal changes in lighting and temperature in Iraq. (From El-Soudi and Al Jebouri, 1979.)

Breed	Annual egg production per bird	Egg weight (g)	Body weight (kg)	Feed consumption g day^{-1}
Native Iraqi	137	49.9	1.61	108
White Leghorn	145	53.8	1.69	112
New Hampshire	115	53.6	2.29	125

Egg production can be most effectively increased by removing the eggs before the birds become broody. This procedure, combined with selection for total egg production, has resulted in many egg-laying breeds capable of producing more than 100 eggs per year. If jungle fowl are kept in captivity and not allowed to incubate, they produce about a third of the eggs produced by a domesticated breed (Hutt, 1949). In a recent study, a native Iraqi breed was found to lay almost as well as a strain of White Leghorn and to lay better than a heavy strain, the New Hampshire (Table 8.1). Such

Table 8.2 A comparison of body weights of wild and domesticated chickens. (From Ensminger, 1980; Nishida, 1980.)

Breed or type	Male (kg)	Female (kg)
Jungle fowl	0.94	0.75
Asiatic breeds		
Brahma (light)	5.44	4.30
Cochin	4.98	3.85
American breeds		
Plymouth Rock	4.30	3.40
Rhode Island Red	3.85	2.94
Wyandotte	3.85	2.95
English breeds		
Sussex	4.08	3.17
Cornish (Dark)	4.53	3.40
Orpington (Buff)	4.53	3.62
Mediterranean		
Leghorn	2.72	2.01
Commercial birds		
White egg-layer	—	1.5–1.8
Brown egg-layer (restricted feeding)		1.7–2.3

native breeds are often more resistant to disease and are better able to withstand high temperatures than White Leghorns. However, their egg production is much inferior to that of commercial hybrid crosses. Domestication has increased the body size of most breeds kept for egg production or for meat (Table 8.2). Part of this increase may have been caused by selection for egg size. Commercially, this trend is currently being reversed by selecting for reduced body size without decreasing egg mass.

Poultry production began to be organized on a commercial basis towards the end of the 19th century, particularly in the United States and Western Europe. During the first three decades of the 20th century nationally organized egg-laying competitions did much to stimulate selection within existing breeds and, as a result, average egg production per hen increased. A progressive increase in fundamental knowledge of poultry genetics, diseases, nutrition, general management and physiology during this period was finally applied in a total package in a commercial context from the 1930s onwards, with spectacular results (Figure 8.1). It is impossible to quantify the relative importance of the contributions of geneticists, nutritionists, pathologists and physiologists to the progressive increase in the average numbers of eggs produced by commercial hens over the last 50 years. Physiologists have contributed most to the management

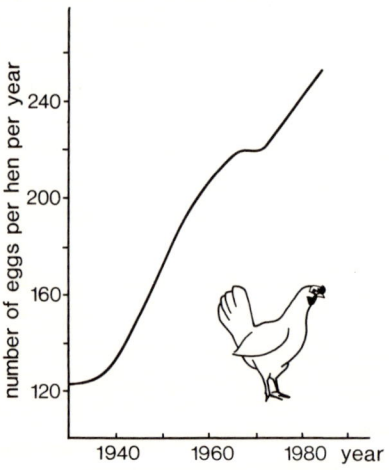

Figure 8.1 The effects of genetic selection and improved nutrition, disease control and management techniques on the average number of eggs produced per hen per year. (Updated from Ensminger, 1980.)

of laying flocks and in particular in the use of lighting patterns (see section 8.2.1) and of controlled temperature environments. By maintaining hens between 13 and 21°C with a relative humidity of 50–75%, the efficiency at which feed is converted to eggs is optimized. A slightly higher temperature, 24°C, is required for the most efficient production of meat.

The discovery by plant breeders that superior strains of maize can be obtained by crossing inbred lines had a great impact on poultry geneticists. This effect, known as hybrid vigour or heterosis, is applied in the production of all commercial egg-layers. Inbred lines, strains or breeds are combined in 3- or 4-way crosses to produce the final commercial product. This inbred stock, the so-called 'grandparent' lines, forms the genetic basis of a commercial cross and is a breeding company's most valuable asset. Consumer preference for either white or brown eggs is met by specially developed brown- or white-egg-laying strains. In 1984, well managed flocks of brown- and white-egg-layers are expected to produce 270 or 279 eggs per bird in a year, respectively. The brown-egg-layers can produce 1 kg of eggs from 2.49–2.56 kg of food while the white-egg-layers can do the same on only 2.32–2.46 kg of food.

In the 1950s chickens began to be bred exclusively for egg production or for meat. The meat-type birds or broilers, like commercial egg-layers, are the result of four or more crosses of parent and grandparent lines. The introduction of the broiler resulted in dual-purpose birds such as the New Hampshire, kept for both egg and meat production, becoming commercially unimportant. Intensive selection for increased growth rate has produced spectacular results (Table 8.3). In the last 30 years efficiency of conversion of feed to meat has doubled while the number of days required to reach a body weight of 2 kg has been halved. Physiologically, a price is paid for these dramatic changes. The parents of the final commercial cross become obese unless their feed intake is severely restricted, and this, in turn, leads to a reduction in egg production. Thus, a well-managed broiler breeder produces only 168–172 eggs in a laying year.

Table 8.3 Effects of genetic selection on the efficiency of broiler production.

	Year		
	1952	1978	1983
Days taken to reach 2 kg live weight	91	56	49
Kg of feed required to produce one kg live body weight	4.4	2.1	2.1

8.1.2 Turkeys

The ancestors of domestic turkeys are native to North America and Mexico and were partly domesticated by the Aztec Indians of Mexico before the Spanish conquest. These birds were first introduced into Europe by the Spaniards in the early 16th century and various European domestic strains developed from them. These strains were subsequently brought to North America by the early European colonists and crossed with native turkeys to produce the ancestors of the present commercial bird.

The turkey is a temperate and sub-tropical zone long-day breeder and has evolved to breed at higher latitudes than the jungle fowl. As a result, its breeding cycle is much more strictly dependent on daylength. For example, the onset of lay in hens reared on short days is not much delayed if they are not exposed to increasing daylengths after they have reached adult body weight. In contrast, the onset of lay in turkeys reared on short days is greatly delayed if they do not receive the appropriate photostimulation. In common with many other temperate zone breeders (Chapter 7), seasonal breeding in the turkey is terminated by the development of absolute photorefractoriness. Thus, in turkeys, selection for increased egg production is more difficult than in the jungle fowl because of the additional physiological barrier imposed by the development of this condition. However, since the wild turkey comes from the lower latitudes in the temperate zone, a prolonged breeding season is possible, with two broods being produced in a season. The first clutch to be produced contains 15–25 eggs and the second 10–15 eggs (Schorger, 1966). The wild turkey thus has the potential to produce well in excess of 40 eggs in the breeding season, especially if the bird is not allowed to complete its clutch. Commercially, turkey breeder hens are expected to produce 70–100 eggs, depending on the strain, before the development of photorefractoriness. However, in some lighter-bodied lines selected, in part, for egg production, hens are capable of laying up to 200 eggs in a laying year, and in these birds, the development of absolute photorefractoriness is not so clearly expressed.

Commercial turkeys that are produced for meat are the result of hybrid crosses between male lines selected for growth rate and female lines selected, in part, for egg production. Different commercial crosses are produced to meet the market requirement for small (4–6 kg), medium (5–12 kg) and large (> 15 kg) birds. Selection for increased rate of body growth has resulted in the large-type commercial birds being nearly three times the weight of their wild ancestors (Figure 8.2).

The effort made to increase the rate of growth has not been paralleled by

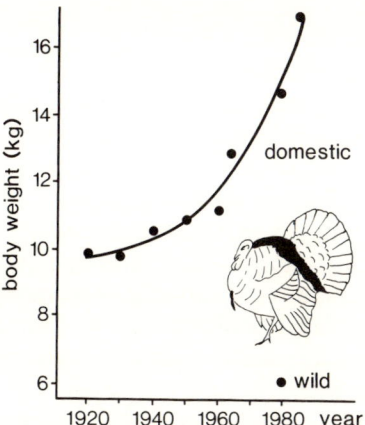

Figure 8.2 The effects of genetic selection and improved nutrition, disease control and management techniques on the average weight of heavy-type male turkeys at 26 weeks of age. (Updated from Ensminger, 1980.)

a similar attempt to increase reproductive performance. This is because simultaneous selection for growth rate and reproductive performance slows down the trend towards increased body weights. The reproductive physiologist is therefore presented with two problems in the turkey which have not yet been fully tackled by the geneticists. These are:

1. How should breeding turkey hens be managed to prevent or delay the development of photorefractoriness?
2. What methods can be used to prevent a loss of egg production through the development of broodiness?

Selection for growth rate also presents a further problem for the reproductive physiologist. The breeding male is so heavy and ungainly that natural mating is impossible. Consequently, commercial turkeys must all be produced by artificial insemination (AI). Successful AI programmes depend on the development of suitable diluents for spermatozoa which requires a knowledge of semen physiology.

8.1.3 *Ducks and geese*

The Mallard is widely distributed in the temperate zone of both New and Old Worlds. It breeds in the spring, and breeding is terminated by the development of absolute photorefractoriness (Chapter 7). Mallards were domesticated in China 2000 years ago and many varieties have been developed for meat or egg production. The best known meat-type breeds are the Aylesbury and Pekin. These birds have been intensely selected to produce ducklings which, depending on the strain, have a live body weight at 47 days of age ranging between 1.29 and 2.97 kg. A further meat-type duck, the Muscovy, is descended from a South American duck, unrelated to the Mallard. The most common egg-producing strains of Mallard are Khaki Campbells and Indian Runners; both have a better rate of lay than the best commercial chicken hybrids. Commercial Khaki Campbells lay 320–340 eggs in their first year, while commercial meat-type ducks lay about 150–250 eggs.

In terms of reproductive performance, domestication has gone further in the duck than in either the chicken or the turkey. There are no known physiological reasons why the reproductive performance of the Khaki Campbell can not also be achieved in other domesticated poultry. The effects of domestication on the endocrine system in poultry are known in detail only in the Khaki Campbell duck (Haase and Donham, 1980). Domestication has resulted in an extension of the breeding season and the disappearance of, or at least a marked reduction in, the period of absolute photorefractoriness. Thus, Khaki Campbell ducks exposed to natural lighting at 54°N lay from February until November with some laying in December and January during mild winters. Further, the birds do not become broody as readily as their wild ancestors. Khaki Campbell ducks usually start incubating only after the number of eggs in the nest is so large that the bird can not cover them all. Domestication also affects the sexual behaviour of the bird. Wild Mallards form firm pair bonds whereas the pair bonds in the Khaki Campbell are weak. Consequently, female Khaki Campbells are more promiscuous than their wild relatives.

In the Khaki Campbell drake, domestication has not only altered the neuroendocrine mechanisms controlling photorefractoriness but has also affected the growth of the testes. The fully developed testes of the Khaki Campbell is about 8 times as large as that of the wild drake and never regresses in the winter, to the same extent as in the wild bird (Table 8.4). This increase in testicular size is not reflected in an increase in concentrations of plasma LH but is related to an increase in the concentration of plasma testosterone.

Domestic geese are descended either from the Greylag Goose or the Swan Goose. Both wild ancestors are temperate zone breeders with a period of absolute photorefractoriness in their annual breeding cycle. Geese are produced for meat, and male meat-producing lines are crossed with lighter

Table 8.4 Seasonal maximum and minimum testicular weights and plasma hormone concentrations in wild and domesticated ducks exposed to natural lighting. (From Haase and Donham, 1980.)

	Testes weight (g)		Plasma LH (ng ml^{-1})		Plasma testosterone (ng ml^{-1})	
	Max.	Min.	Max.	Min.	Max.	Min
Wild	19.9	0.32	3.43	0.32	3.80	0.02
Domestic (Khaki Campbell)	159	15	3.14	0.68	5.94	0.44

female egg-producing lines to produce a commercial hybrid. Economically, goose production is more important in the Soviet bloc than in the West (Saleyev, 1975).

8.2 Intensive farming

Poultry are produced under intensive farming conditions to improve the efficiency of meat and egg production. Reproductive physiologists contribute in several ways to improving the management of poultry under intensive conditions, particularly in relation to lighting patterns, the control of broodiness and the induction of moult. The maintenance of poultry in intensive conditions raises ethical problems, and here too the reproductive physiologist can contribute to the debate on animal welfare.

8.2.1 *Lighting patterns*

The beneficial effects of artificial lighting on egg production in chickens was noted as long ago as 1889. At that time it was thought that extra light at night stimulated laying by encouraging the birds to eat the feed required for egg formation. When chickens are held under intensive conditions exposed to natural lighting, the scope for controlled lighting is limited. The standard practice is to supplement the prevailing daylength so that it equals about 14 hours or the longest daylength in the year, whichever is the greater. In this way, the birds are held on a constant long day. This lighting treatment has two desirable effects. First, it prevents a seasonal drop in egg production during the autumn and winter, and secondly, it delays the onset of lay. This is important, because if hens are reared from an early age on an increasing daylength, egg production is stimulated before they are fully grown. The eggs laid are too small for commercial use and the birds' subsequent rate of lay is poor. The delaying effect of long days on sexual maturation is due to the development of juvenile photorefractoriness. As discussed in section 7.1, this is a mechanism found in many long-day breeders which prevents young from breeding in the year in which they are hatched.

Unlike the chicken, in the turkey juvenile photorefractoriness is not readily dissipated after prolonged exposure to long days. Somatically mature turkey hens require exposure to at least four weeks of short days to dissipate juvenile photorefractoriness before they can be brought into lay by increasing the daylength.

At higher temperate-zone latitudes, poultry are housed most economically in windowless, well-insulated, buildings in which the temperature and lighting are artificially controlled. Under these conditions chickens are reared on short days to dissipate juvenile photorefractoriness and to prevent sexual maturation until the appropriate body size is reached. This is usually when the birds are between 16 and 22 weeks old, depending on the strain. At this time the daylength is increased in half-hourly steps at weekly intervals. The critical daylength in the domestic hen is between 10 and 11 hours, and a gradual increase in daylength above the critical daylength allows gonadotrophin secretion to increase slowly and the reproductive organs sufficient time to develop properly. Commercial chickens treated in this way are brought uniformly into lay at the earliest age at which the size of the eggs laid is commercially acceptable. The laying performance of a flock of commercial chickens over a laying year is directly related to the peak of egg production achieved after they are first photostimulated. This peak, which in commercial layers is about 95% of the total number of eggs theoretically possible, depends in part on photostimulation of the birds in the way just described.

The different ways in which lighting patterns are used to manipulate egg production and quality in commercial egg-laying chickens are shown in Table 8.5. Egg size can be increased and shell quality improved by entraining the ovulatory cycle of laying hens to lighting cycles greater than 24 hours. The 'open period' of the ovulatory cycle (section 6.5) occurs once during each lighting cycle, and hence the interval between successive ovulations is increased. This allows extra time for more yolk deposition in the largest ovarian follicle, which ultimately results in a larger egg. The lengthened ovulatory cycle also increases the time the egg is in the shell gland (section 6.1) and hence improves shell calcification. Intermittent–symmetrical lighting schedules have a similar effect on shell quality and egg size, but the precise physiological reasons are uncertain. Additionally, by

Table 8.5 Lighting pattern used to regulate egg production and quality in commercial egg-laying chickens.

Type of lighting pattern	Benefits	Disadvantages
Ahemoral e.g. 28-hour cycle of 14 h light, 14 h darkness (14L:14D)	Increased egg size e.g. 59.7 → 63.3 g Improved shell quality Saves electricity	Reduced egg number
Intermittent–symmetrical e.g. 6L:6D:6L:6D	Bigger eggs Saves electricity Better shell quality Better feed conversion	Reduced egg number
Intermittent–asymmetrical e.g. 2L:4D:8L:10D (Cornell Pattern)	Saves electricity	None
Biomittent e.g. (0.25L:0.75D) × 16:8D	Saves electricity Better feed conversion	Possible reduction in egg size Inconvenient working conditions

stimulating feeding at short regular intervals the efficiency of feed conversion to eggs is increased. Intermittent and asymmetrical lighting schedules were developed as a result of research on the role of circadian rhythms in the control of photoinduced gonadal growth (section 7.3). The realization that photostimulation depends not on how much light a bird receives in 24 hours, but where it occurs, led to the widespread use of 8-hour light period during the working day, preceded by a 2-hour period of light in the middle of the night. The resulting saving in lighting costs can be economically important in well-managed flocks operating on a tight profit margin. A commercial lighting pattern called 'biomittent lighting' involves the provision of light for 15 minutes in every hour for 16 hours each day. This system of lighting is not based on any known physiological effects of light but also saves on lighting costs and improves the efficiency of the conversion of feed to eggs. Lighting patterns of the type shown in Figure 8.5 are not widely used in poultry other than chickens.

The control of lighting is particularly important in achieving the best egg production from turkey breeders. It is customary to rear birds on short days and to increase the daylength to 14 hours at 30 weeks of age. In view of the studies of the role of light in the development of photorefractoriness in wild birds (section 7.7) it is likely that 14 hours of light serves to stimulate gonadotrophin secretion without driving the bird too rapidly into photorefractoriness.

8.2.2 Broodiness

The term 'broody' is used to describe both the incubation and brooding phases of parental behaviour (section 6.6). Since the reproductive system of broody birds regresses, broodiness from a commercial point of view is a cause of lost potential egg production. Commercial chickens and turkeys are all hatched from artificially-incubated eggs, and consequently, the broody trait is commercially unwelcome. Genetic selection and management techniques are therefore used to suppress broodiness in breeding stocks and commercial hybrid egg layers. The onset of incubation behaviour is stimulated by several environmental stimuli of which the most important are the sight of the nest and the presence of an accumulating clutch. Such stimuli are readily removed by keeping hens in cages. The maintenance of laying hens in this way, combined with genetic selection, has virtually eliminated broodiness in commercial hybrid egg layers. However, the situation in breeding stocks is different since they are generally kept in floor pens. Under these conditions, the environmental stimuli which stimulate the onset of incubation are more likely to be present. Lost egg production due to the development of broodiness is therefore a problem in some breeding flocks and in particular in the heavy strains of turkeys. In these birds, selection is essentially for growth rate, and selection against broodiness is much less than in other types of commercial breeding stock. The depressive effects of broodiness on egg production in a flock of turkey breeders are illustrated in Figure 8.3. Since the onset of incubation behaviour is preceded by an increase in nest visits and in the

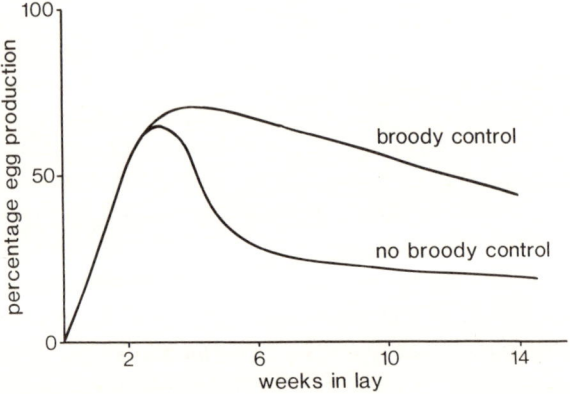

Figure 8.3 Egg production in a flock of heavy-type Turkey breeders showing the effects of broodiness on egg production.

time spent on the nest (section 6.6), good management of commercial breeding flocks depends on the stockman being able to identify this pre-incubation behaviour. In addition, broodiness is discouraged by regularly rotating laying hens between different pens. In this way, the birds are deprived of a familiar nest site, and one of the stimuli which induce incubation.

Once a bird has become broody and has stopped laying, it is treated to bring it back into lay as quickly as possible. The treatment usually involves isolating the bird in an unfamiliar environment. Treatment with gonadal steroids also terminates broodiness but it is not more effective in bringing the birds back into lay than most non-hormonal treatments. Research on the neuroendocrine control of gonadotrophin and prolactin release (sections 6.2, 6.6) may lead to the development of therapeutic methods, using neuropeptides and neurotransmitters, to terminate broodiness and stimulate a rapid resumption of egg production.

8.2.3 *Induced moult*

As commercial layers age, egg production falls and eggshell quality deteriorates. These changes are not irreversible because egg production and shell quality improves if the reproductive organs are induced to regress and subsequently to redevelop. This is achieved by inducing a moult, thereby mimicking the endocrine changes which occur at the end of a breeding season in wild birds (section 6.7).

Induced moulting is standard husbandry practice in many commercial hybrid egg-laying flocks and effectively extends their productive laying life. Knowledge of the physiology of moult has been applied to develop several effective methods to induce moult (Wolford, 1984). The objective of all these methods is to induce ovarian regression, and consequently the cessation of lay, within 8 or 9 days and to cause a loss in 20–25 percentage of body weight in about 14 days. The loss in body weight is important since it reduces excessive body fat which has an adverse effect on egg production. The methods used for induced moulting include feed and/or water withdrawal, light restriction and the feeding of diets low in calcium or sodium or high in zinc or iodine.

8.2.4 *Stocking density and reproductive performance*

Chickens and turkeys, in common with other galliform species, do not require 'essential supplementary' proximate information to stimulate the

Table 8.6 Stocking densities of commercial breeding stocks, not held in cages. (From Ensminger, 1980.)

Type of bird	Stocking density (cm² per bird)
Broilers	1858–2322
Commercial layers (Leghorn-type hybrids)	1394
Turkeys (large-type)	2322–4181
Ducks	2322
Geese	2322

final growth of yellow-yolky ovarian follicles (section 7.2). In this respect, chickens and turkeys differ from non-galliform species, making it possible for eggs to be obtained from them at high stocking densities. Most breeding stocks of poultry are kept in floor pens where stocking densities range between 1400 cm² per bird for light-bodied commercial birds to 4200 cm² per bird for heavy-bodied turkeys (Table 8.6). Commercial hybrid chickens producing infertile eggs for consumption are held at higher stocking densities in battery cage systems. More than 96% of all commercial egg-layers are kept in this way in the USA and Western Europe. It is tempting for egg producers to put as many birds into a cage as possible. However, as stocking density increases there is a reduction in egg production (Hughes, 1975; Figure 8.4). The precise physiological reason for this reduction is unknown but probably involves the same mechanism which inhibits laying in wild birds in stressful situations which occur naturally (section 7.2). As

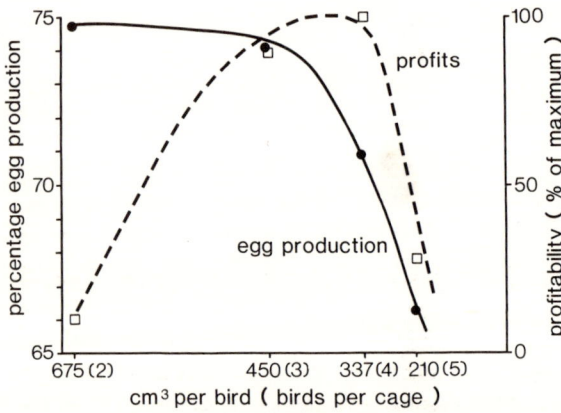

Figure 8.4 The effects of stocking density on egg production and profitability in commercial hybrid egg-layers kept in 30 × 45 cm battery cages. (From Card and Nesheim, 1972.)

APPLIED ASPECTS

Figure 8.5 The minimum floor space as proposed by the EEC Commission to be permitted for caged laying chickens kept in member states of the EEC. A medium hybrid brown egg-layer is shown standing in a 405 cm² square. This is at the lower limit of the tolerance set for the recommended 450 cm² per bird.

shown in Figure 8.4 it can be seen that maximum profits can be gained by stocking birds at densities which are beginning to have a depressive effect on egg production.

The maintenance of chickens for egg production in this way is a matter public concern. Guidelines have been proposed for the minimum floor space to be provided for caged layers and the most recent, produced by the EEC Commission, suggest that each bird should have 450 cm² floor space with a 10% tolerance level (Figure 8.5). This may become a legal requirement in all EEC countries and some Western European countries may ban battery cages completely. This was tried for a few years in Denmark in the 1970s with disastrous consequences for the egg-producing industry while the Swiss have passed legislation banning battery cages from 1991. The future of intensive egg production depends on whether the consumer is prepared to pay more for eggs produced in alternative systems.

8.3 Pollution

The emphasis of this book so far has been on birds in their natural environments. As we have seen, the ability of species to radiate into potentially hostile environments turns on the efficiency of mechanisms to maintain homeostasis. These can be behavioural adjustments which enable the species to avoid the stressor or physiological mechanisms which negate the deleterious impact of the stressor. Examples of both have been described. There is, however, another dimension to the environment which increasingly threatens the survival of ecosystems and therefore of species within them, namely the widespread use of pollutants in agriculture and industry. These include agrochemicals and insecticides on the one hand, and the accidental, non-accidental or inevitable release into the environment of pollutants such as heavy metals, organo-chlorines, hydrocarbons and radionuclides. Perhaps the most dramatic example is when oil spills into the ocean following tanker disasters. Such occurrences, despite the immediate impact on the environment, should not diminish our appreciation of the insidious build-up of pollutants which eventually culminate in morbidity and even mortality. Pollutants enter food chains at various levels but, in general, top predators such as birds are more threatened than, say, invertebrates at the bottom of the food chain. This results from the fact that the volume of food required to sustain energy intake and meet metabolic needs is greatest at this level; put in another way, low and sub-lethal levels of contaminants in organisms, be they animal or plant at the bottom of the food pyramid, provide the potential to become increasingly concentrated in organisms further up the pyramid. The possibility exists that eventually a lethal effect becomes apparent by a process of accumulation. Such applies to all pollutants, but the possibility also exists that more than one pollutant may simultaneously feature to threaten survival of birds. (Having highlighted the problem, it must also be stated that certain persistent chemicals occur in organisms for which there is yet no evidence of any adverse or cumulative effects.) Pollutants therefore exist as a threat to survival. When the threat does not assume lethal proportions, birds tolerate rather than adapt to the changed characteristics of their environment.

Considerable attention has been focused in recent years on the impact of pollutants on man and other animals. The growth of the problem is associated with the expansion of industrialization, a trend which is increasingly gaining strength as technologically advanced countries expand and diversify their manufacturing base and as underdeveloped countries seek to compete in growing world markets.

'Pollutants' is a generic term which covers all substances which are released into the environment and which have a potential adverse effect on animals, plants and the ecosystems which they collectively comprise. Natural pollution does, of course, occur, as for example the seepage of oil into the sea off California, but is a limited threat when compared to the totality of the problem produced as a result of man's activities.

With respect to birds, greatest public attention has been focused on the effect of petroleum contaminants on whole populations. The problem has arisen following the enormously increased transport of crude oil and its byproducts around the world. Despite international recognition of the problem and measures to restrict the illegal disposal of waste at sea, largely through the requirement to ballast empty bulk tankers for their return journeys to oil-producing countries, pollution continues to occur in this way. The problem is hard to contain because of the complex international laws governing oil tankers; the principle of the freedom of the high seas dictates that a vessel is subject to the authority of the state under whose flag it sails. It follows that if such a state disregards international conventions, then its registered vessels can discharge waste with impunity even within areas which are internationally agreed as inviolate. The Torrey Canyon disaster highlights the difficulties of apportioning responsibility for spillage. This vessel was owned by a USA company, registered under a Liberian flag, chartered by a British company with an Italian crew, and foundered in international waters. These disasters, of course, are dramatic in the extent of the destruction of wildlife; they tend to mask, however, the ongoing problems of the incremental build-up of pollutant levels derived from smaller intentional or accidental spillages and dumping of waste. This problem is growing in severity and the extent to which it is occurring has been highlighted by round-the-world yachtsmen who describe 'films of oil' covering most of the open seas of the world.

8.3.1 Physical effects of oil contamination in birds

Total immersion in oil results in rapid mortality of both mature and young animals. Mortality due to lesser levels of contamination does, however, occur. For example, minor contamination of adult feathers can lead to the transference of oil to eggs and fledglings in the nest and lead to death of unhatched chicks and young birds.

Contamination of feathers disrupts the aerodynamic qualities of the body contours as well as directly affecting the mechanical characteristic of the feather barbules. Flight is therefore affected, and in aquatic species buoyancy decreases to a point where birds can drown. Plumage disruption leads to displacement of the entrapped thermo-insulting volume of air, and the penetration of water exacerbates the problem of restricting heat loss. Such events impose added burdens on thermoregulatory homeostasis. To maintain body temperature, additional food becomes a requirement at times when foraging for food becomes increasingly difficult because of the added handicap in impaired flight and the possible addition to body weight due to the decrease in waterproofing of the feather mass. Changes in metabolic rate can lead to dehydration as a result of a higher rate of insensible respiratory water loss, and such may well account for the high incidence of emaciation in birds experiencing only moderate contamination with petroleum.

8.3.2 Effect on the embryo

The transference of even small amounts of oil from the breast feathers to the eggs during incubation results in embryonic mortality. This has been observed in terns and gulls, and experimentally shown in ducks, which failed to produce viable young from incubated clutches. The hatchability of artificially incubated eggs which had been smeared with oil is similarly reduced to less than 20% of controls.

Petroleum products, when applied to eggshells of fertilized eggs, cause increased embryonic death and stunted growth of those chicks which hatch. Their action is twofold, depending on the type of contamination applied to the eggshell: viz. an impairment of across-the-shell respiration with or without a direct toxic effect on the embryo.

8.3.3 Systemic effect of oil pollution

Possible routes for the internalization of oil or its volatile byproducts are by inhalation, via the skin, via the eggshell to affect the tissues of the embryo or via the gut by drinking of contaminated sea-water, swallowing contaminants during preening or by the ingestion of contaminated food. Unequivocal evidence that petroleum products pass through the gut wall is provided by the fact that they accumulate in body tissues and evoke increases in detoxifying liver enzymes. The role of the liver in the metabolism of these pollutants is crucial because it possesses a system whereby hydroxylation of the lipophilic contaminants renders them more water-soluble for rapid excretion from the body. The importance of the liver as a site of metabolism is associated with the mixed function oxidase system (MFO) residing within the liver microsomal elements.

8.3.4 Effect on juveniles

Young birds, especially those immediately post-hatching, seem very sensitive to ingested petroleum. A whole range of physiological parameters are affected including some which are endocrine, principally those related to growth and osmoregulation. Growth rate is impaired, but not for reasons of impaired appetite, for contaminated birds frequently indulge in hyperphagia, and the lesion may occur in an impaired uptake of nutrients through the gut due to pollutant-related pathological changes in the mucosa. The other important physiological change relates to the capacity of young birds to osmoregulate. As mentioned elsewhere (p. 85), the essential first step in salt and water homeostasis is the isosmotic uptake of fluid from the gut contents and the subsequent differential excretion of Na^+ in order to gain osmotically free water for metabolic purposes. The presence of petroleum products in the diet diminishes the high rates of mucosal transfer when ducklings are on a regime of hypertonic drinking water; in fact, the normal adaptational changes in the gut transport mechanisms can be abolished within one day by giving small doses of petroleum. Mortality in juveniles is linked with high Na^+ levels resulting from dehydration rather than an accumulation of plasma Na^+. Given this situation, alleviation would follow through the activation of the extra-renal mechanisms of excreting Na^+. This appears not to be the case, despite the fact that hypertrophy of the salt glands occurs, but this hypertrophy is paralleled by a decrease in the specific activity of the Na^+-K^+-dependent ATPase, essential for the secretory processes.

8.3.5 Responses in the endocrine system

The changes already observed would suggest a significant endocrine response. This could be of two kinds; either changed level of production and release of key hormones, or a changed response of the target organs to the humoral factors. Certainly, corticosterone and thyroxine levels are elevated and remain at a high level and then decrease, a situation reminiscent of the alarm reaction of Selye's General Adaptation Syndrome. The probable explanation is summarized by Holmes (1984) as follows.

> 'The experimental evidence suggests that a syndrome comprised of two simultaneous sequelae may occur when juvenile birds ingest small amounts of petroleum. The first resembles a typical response to acute stress, characterized initially by a stimulation of the pituitary–adrenal axis and later by stimulation of the pituitary–thyroid axis. At the same time, petroleum compounds present in the gut appear to compete with endogenous corticosteroids for binding sites in the responsive cells of the intestinal mucosa, to prevent the maintenance of normal water, and perhaps also nutrient, uptake from the gut. The net effect of the hyperadrenocortical state and the simultaneous acute dysfunction in the corticosteroid-sensitive mucosal cells, could fully account for the gross physiological changes that are seen in contaminated juvenile birds'.

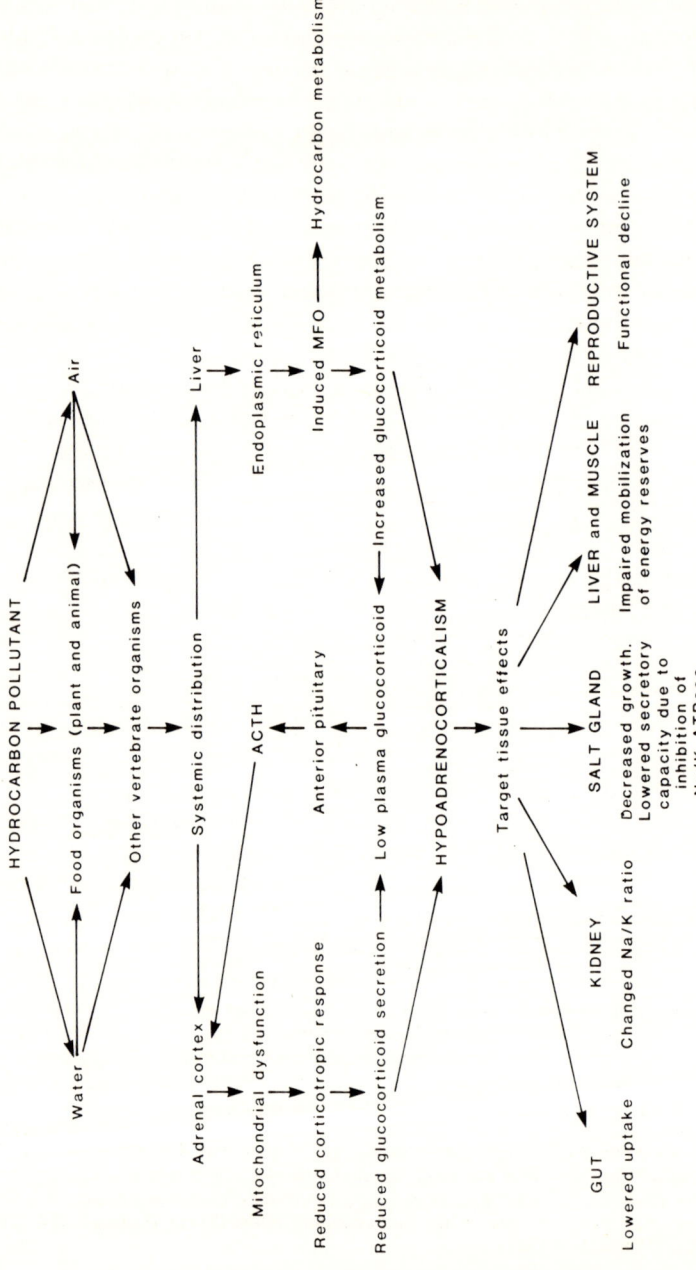

Figure 8.6 The pathways whereby petroleum hydrocarbon pollutants may affect adrenocortical function in birds and the consequent effects on regulatory processes. (Modified from Homes *et al.*, 1981.)

APPLIED ASPECTS   193

Growth and development with its consequent metabolic demands might be considered a more vulnerable period in the life-history. Indeed, adult birds are capable of tolerating petroleum pollution more successfully than juveniles. In fact, birds given adequate protection from extremes of environment display little, if any, discomfiture from petroleum ingestion. What is clear, however, is that such birds are distinctly more vulnerable to environmental stresses such as cold, hypersalinity and crowding, and the common denominator in terms of environmental response lies in a changed level of activity of the pituitary–adrenocortical axis; such a situation has profound physiological effects and the scheme describing the resultant sequelae is outlined in Figure 8.6.

8.3.6 *Impact of petroleum products on reproductive performance*

Exposure of breeding animals to petroleum products, either in experimental procedures or as a result of exposure under natural conditions, results in widespread and deleterious effects covering all aspects of the reproductive

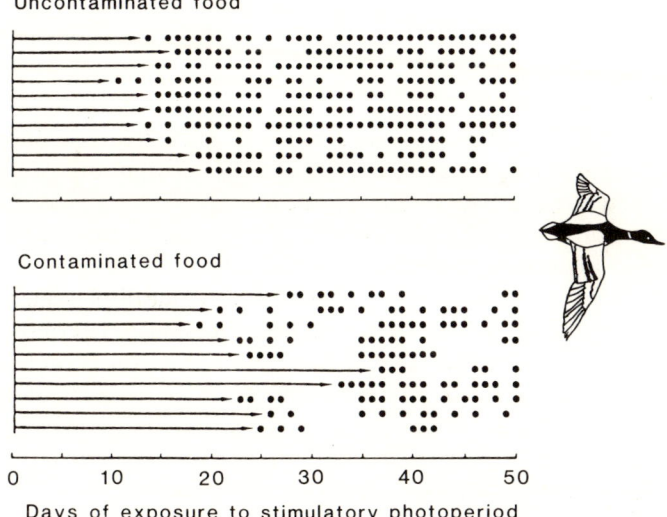

Figure 8.7 The typical patterns of egg-laying by individual unmated female Mallard ducks consuming uncontaminated food and food contaminated with 3 ml South Lousiana crude oil per 100 g dry wt. The birds were exposed to a light regime on day zero and at the same time crude oil was added to the food of the contaminated birds. (From Holmes, 1984, after Cavanaugh and Holmes, 1982.)

process. These include a retardation in oocyte formation and ovarian differentiation, disturbed oviposition, decreased fecundity, decrease in the viability of the hatched eggs and an impairment of shell formation. Added to which reproductive behaviour becomes abnormal and there is also evidence of an attenuation of the reproductive cycle. The evidence for this is based on experiments using ducklings and Japanese Quail. In general terms these perturbations in reproductive behaviour and performance can be reversed when the pollutant stressor is removed (see review by Holmes, 1984), but the physiological lesion induced by exposure to petroleum products may persist after the return to normal feeding regimes, as described in experiments on domestic ducks. The observed changes in reproductive parameters would *a priori* be expected to have an underlying endocrinological cause. Plasma oestradiol and oestrone levels increase during ovarian maturation at a lower rate in contaminated birds than in controls, and a similar, though less marked, pattern applies also to peripheral progesterone. The evidence would suggest that the gonadal dysfunction is not as a result of impaired gonadotropin secretion, an observation which suggests a direct effect on the germinal components. The expression of these effects in terms of changed patterns of egg-laying is shown in Figure 8.7.

8.3.7 *Incubation and breeding*

Contaminated birds display irregular cycles characterized by a protracted phase of gonadal development and irregularity of oviposition with many females failing to incubate the clutch; the overall effect is a lengthening of the reproductive cycle by up to 50%. The central controlling role of prolactin in the behavioural and physiological events required to initiate and sustain the cycle has been well established, and the displacement of the pattern of reproduction in contaminated birds just described in all probability reflects the phase shifts in secretion and release of prolactin, for LH remains unaffected except in the early part of the ovarian maturation phase.

8.3.8 *Embryonic and post-embryonic effect of pollutants*

Following exposure to oil or its by-products, the MFO levels rise markedly, as does liver size, and the liver must be regarded as a major site of detoxification. Evidence also points to accumulation in other tissues including egg yolk. When the yolk is taken up by the developing embryo, an increase in MFO in the embryo's liver results, similar to that found in the parent. An alternative route of contamination might be via the eggshell from pollutants carried on the parent's feathers. This is unlikely because artificially incubated eggs treated externally with pollutant fail to evoke a similar response in the developing embryo. It is clear, however, that the toxic effect of pollutants which persist into the hatchling stage increases their vulnerability to adverse environmental conditions.

8.3.9 *Relevance of laboratory studies to studies in the wild*

The basic rationale of all studies in the field of avian physiology is to arrive at a better understanding of evolved strategies which secure the success of the species. In this context, thought must be given to the way laboratory studies can be extrapolated into the ecosystem of a particular species. Holmes (1984) has recently addressed this problem.

Laboratory studies clearly indicate that contamination of birds renders the individual more susceptible to homeostatic failure. On the wider scale whole populations can be threatened. This is particularly so for species living in vast colonies utilizing feeding grounds which coincide with major shipping routes. Pollution of the environment would result in contamination of animals and plants in the food chain. Some of the bird colonies living along the coastlines of the continental shelf consume vast amounts of food. One report quotes a figure of more than 20% of the annual production of pelagic fish. It follows, therefore, that there is a good chance of sub-lethal levels of pollutants building up in top predators. In the case of birds this leads to impaired efficiency as a result of physical contamination and to an impairment of metabolism when adrenocortical insufficiency follows ingestion of hydrocarbons. Taken together, these effects will subtract from the ability of birds to sustain active flight in the quest for food. Nestlings suffer too because many species feed their young with regurgitated food.

Some species such as murres, puffins, auks and razorbills are long-lived, reaching maturity late in life, but their low rates of reproduction result in low rates of recruitment to the population. It has been reported that recruitment rates can be as low as 0.2–0.6 fledglings per breeding pair per year. It is clear that an increase in mortality rate would have a catastrophic effect on whole populations and in their capacity to subsequently restore colony numbers. There is some evidence from demographic studies on colonies of two species of murre and two species of kittiwake inhabiting the

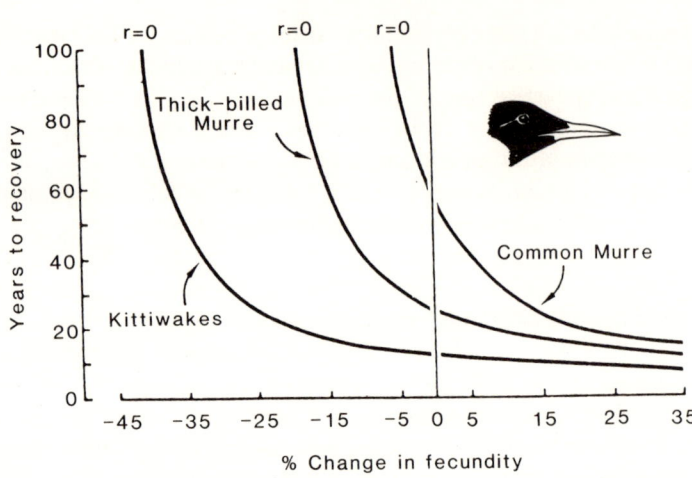

Figure 8.8 The calculated effect of sustained reductions in fecundity on the restoration of some sea bird populations following a single episode involving 50% mortality among all age-classes (adults and juveniles). r = coefficient of annual recruitment to the population. (From Holmes, 1984, adapted from data of Weines et al., 1982.)

Pribilof Islands in the Bering Sea (see Holmes, 1984). The life-history of each species was examined and a model constructed to describe the recovery patterns that might follow when populations were depleted by single episodes of mortality. Theoretically it would take a colony of kittiwakes some 10 years to recover from an episode of mortality involving 50% of the birds, whereas the two species of murres would require a longer (25 and 60 years) recovery period (Figure 8.8 vertical line on zero change in fecundity). Additionally, the remaining population could suffer reduced fecundity because of impairment of the hormonally-regulated reproductive process. Examination of the examples given in Figure 8.8 shows that, in the case of the Common Murre, a fall in fecundity of only 5% would be sufficient to reduce recruitment (r) to zero and extinction of the colony would follow. Comparable figures for the Thick-billed Murre and the kittiwakes are 20 and 40% respectively (Holmes, 1984).

8.3.10 Other pollutants

The worldwide demand for the control of pests has resulted in an ever-increasing pressure on the agrochemical industry to produce chemicals with proven efficaceous properties. Substances are released into the environment as a result of manufacturing processes; some of these are persistent and can be toxic. Also, heavy metals are commonly found in the environment as a by-product of various manufacturing processes. These pollutants can enter food chains at various levels, and, for birds, can be ingested directly or as part of a dietary component.

A measure of the scale of the problem is illustrated by a simple statistic: in the last 50 years, more than one million metric tonnes of polychlorinated biphenyls (PCB) have been released into and persist in nature. This worldwide pollutant, which is structurally similar to DDT, has adverse effects, because of its biological stability and its capacity to accumulate in the fatty tissues and food chains, in species such as birds of prey. Experiments using 14C labelled PCB indicate a preferential accumulation in steroid-producing endocrine tissues such as the gonads and the adrenals as well as the fat bodies of both male and female birds, including, importantly, the yolk. Thus, the effect of these contaminants can spill over into the next generation. Moreover, a significant route of excretion is via the eggs — apparently this is a general mechanism of excretion of lipophilic pollutants in birds. Thus, the attendant contamination of embryos involving mortality, decreased hatchability and teratogenic effects is not surprising. It follows that PCB along with other pollutants can lead to smaller clutch sizes and decreased breeding success in wild birds, as hypothesized in the raptor-pesticide-syndrome promulgated by Ratcliffe. Reproduction apart, PCBs show a strong affinity to adrenocortical tissue, and, like DDT may reduce adrenocortical function either directly or as a result of altered thyroid activity. The consequences of this are apparent when the importance of adrenocortical hormones is assessed in relation to osmoregulation and stress resistance.

Pesticides such as DDT also affect eggshell thickness due to the inhibition of carbonic anhydrase, the enzyme responsible for making calcium available to the eggshell in the oviduct. Further, they cause induction of enzymes in the liver responsible for breakdown of steroid hormones, depressing the levels of blood oestradiol and allowing, as a consequence, the mobilization of bone calcium and the observed alteration in the breeding cycle. The thinning of the eggshell leads to a fragile condition and can cause heavy losses during incubation.

The build-up in the widespread use of DDT and related compounds resulted from the fact that DDT is virtually non-toxic in man, and because of this and its efficacy as an insecticide was considered as an ideal control agent. Many years later the true effect of DDT became apparent as a result of its insidious side effects on fisheries and wildlife. One dramatic example of the impact of organochlorine pesticide on bird population is documented by Newton and Haas (1984). Around 1960 there was a population crash in the Sparrowhawk in the United Kingdom which could be correlated in its severity with the proportion of tilled land in the particular region, and hence the usage of pesticides. Recovery of the population has followed successive restriction orders on the use of insecticides. The pattern of recovery has taken place in a wave-like fashion from west to east, occurring first in areas with least-tilled land and where, of course, there had been the smallest decline in the population.

The impact of insecticides on Sparrowhawks included effects of liver accumulation of insecticide residues, reduced breeding success, increased eggshell thinning and increased adult mortality. The recovery of populations took place despite the persistence of insecticides in the environment, but the fall-off in insecticide usage allowed breeding success to be restored to a level where recruitment exceeded mortality due to all causes.

REFERENCES

Able, K.P. (1980) Mechanisms of orientation, navigation and homing. In *Animal Migration, Orientation and Navigation* (Gauthreaux, S.A., ed.), Academic Press, New York, 283–373.

Able, K.P. (1982) The effects of overcast skies on the orientation of free-flying nocturnal migrants. In *Avian Navigation* (Papi, F. and Wallraff, H.G. eds.), Springer, Berlin, 38–49.

Adkins, E.K. (1978) Sex steroids and the differentiation of avian reproductive behaviour. *Am. Zool.* **18**, 501–509.

Adkins, E.K., Boop, J.J., Koutnik, D.L., Morris, J.B. and Pniewski, E.E. (1980) Further evidence that androgen aromatization is essential for the activation of copulation in male quail. *Physiol Behav.* **24**, 441–446.

Adkins, E.K. and Pniewski, E.E. (1978) Control of reproductive behaviour by sex steroids in male quail. *J. Comp. Physiol. Psychol.* **92**, 1169–1178.

Adkins-Regan, E. (1981). Effect of sex steroids on reproductive behaviour of castrated male ring doves (*Steptopelia* sp.). *Physiol and Behav.* **26**, 561–565.

Aidley, D.J. (1981) Questions about migration. In *Animal Migration* (Aidley, D.J., ed.) Cambridge University Press, Cambridge, 1–8.

Alerstam, T. (1981). The course and timing of migration. In *Animal Migration* (Aidley, D.J., ed.), Cambridge University Press, Cambridge, 11–53.

Alerstam, T. and Högstedt, G. (1983). The role of the geomagnetic field in the development of birds' compass sense. *Nature* **306**, 465–465.

Amanova, M. (1975) Adaptive features of the relief structure of the intestinal mucous membrane in desert birds. *Dokl. Akad. Nauk. S.S.S.R.* **225**, 508–510.

Assenmacher, I. (1974) External and internal components of the mechanism controlling reproductive cycles in drakes. In. *Circannual Clocks* (Pengelley, E.T., ed.), Academic Press, London, 197–248.

Bahr, J.M., Wang, S.C., Huang, M.Y. and Calvo, F.O. (1983) Steroid concentrations in isolated theca and granulosa layers of pre-ovulatory follicles during the ovulatory cycle of the domestic hen. *Biol. Reprod.* **29**, 326–334.

Barfield, R.J. (1971) Activation of sexual and aggressive behaviour by androgen implanted into the male ring dove brain. *Endocrinol.* **89**, 1470–1476.

Bastian, J.W. and Zarrow, M.X. (1965) A new hypothesis for the asynchronous ovulatory cycle of the domestic hen. *Poult. Sci.* **34**, 776–788.

Bartholomew, G.A. and Cade, T.J. (1956) Water consumption of house finches. *Condor* **58**, 406–412.

Bartholomew, G.A. and Cade, T.J. (1963) The water economy of land birds. *Auk* **80**, 504–539.

Baudinette, R.V., Loveridge, J.P., Wilson, K.J., Mills, C.D. and Schmidt-Nielsen, K. (1976) Heat loss from feet of herring gulls at rest and during flight. *Am. J. Physiol.* **230**, 920–924.

Baudinette, R.V., Tonkin, A.L., Orbach, J., Seymour, R.S. and Wheldrake, J.F. (1982) Cardiovascular function during treadmill exercise in the turkey. *Comp. Biochem. Physiol.* **72A**, 327–332.

Bech, C., Johansen, K., Brent, R. and Nicol, S. (1984). Ventilatory and circulatory changes during cold exposure in the Pekin duck *Anas platyrhynchos*. *Resp. Physiol.* **57**, 103–112.

Bech, C., Rautenberg. W., May, B. and Johansen, K. (1982) Regional blood flow changes in response to thermal stimulation of the brain and spinal cord in the Pekin duck. *J. comp. Physiol.* **147**, 71–77.

Berger, M. and Hart, J.S. (1974) Physiology and energetics of flight. In *Avian Biology*, Vol. IV, Farner, D.S. and King, J.R. (eds.), Academic Press, New York, 415–477.

Berger, M., Hart, J.S. and Roy, O.Z. (1971) Respiratory water and heat loss of the black duck during flight at different ambient temperatures. *Can. J. Zool.* **49**, 767–774.

Bernstein, M.H. (1976) Ventilation and respiratory evaporation in the flying crow, *Corvus ossifragus*. *Resp. Physiol.* **26**, 371–382.

Bernstein, M.H., Curtis, M.B. and Hudson, D.M. (1979) Independence of brain and body temperatures in flying American kestrels, *Falco sparverius*. *Am. J. Physiol.* **237**, R58–R62.

Berthold, P. (1975) Control and metabolic physiology. In *Avian Biology*, Vol. V, Farner, D.S. and King, J.R. (eds.), Academic Press, London, 77–124.

Black, C.P. and Tenney, S.M. (1980) Oxygen transport during progressive hypoxia in high-altitude and sea-level water fowl. *Resp. Physiol.* **39**, 217–239.

Bono-Gallo, A., Licht, P. and Papkoff, H. (1983) Biological and binding activities of pituitary hormones from the ostrich *Struthio camelus*. *Gen. Comp. Endocrinol.* **51**, 50–60.

Brackenbury, J.H. and Gleeson, M. (1983) Effects of P_{CO_2} on respiratory pattern during thermal and exercise hyperventilation in domestic fowl. *Resp. Physiol.* **54**, 109–119.

Braithwaite, L.W. (1976) Environment and timing of reproduction and flightlessness in two species of Australian ducks. In *Proc. 16th Internat. Ornithol. Congr.*, Aust. Acad. Sci., Canberra, 489–501.

Braun, E.J. (1978) Renal response of the starling (*Sturnus vulgaris*) to an intravenous salt load. *Am. J. Physiol.* **234**, F270–F278.

Braun, E.J. (1982) Renal function. *Comp. Biochem. Physiol.* **71A**, 511–518.

Braun, E.J. and Dantzler, W.H. (1972) Function of mammalian type and reptilian type nephrons in the kidney of the desert quail. *Am. J. Physiol.* **222**, 617–629.

Brent, R., Pedersen, P.F., Bech, C. and Johansen, K. (1984) Lung ventilation and temperature regulation in the European coot *Fulica atra*. *Physiol. Zool.* **57**, 19–25.

Brodsky, L.M. and Weatherhead, P.J. (1984) Behavioural thermoregulation in wintering black ducks: roosting and resting. *Can. J. Zool.* **62**, 1223–1226.

Butler, P.J. (1981) Respiration during flight. In *Advances in Physiological Sciences*, Vol. 10, Hutás, I. and Debreczeni, L.A. (eds.), Pergamon Press, Oxford, 155–164.

Butler, P.J. (1982) Respiration during flight and diving in birds. In *Exogenous and Endogenous Influences on Metabolic and Neural Control*, Addink, A.D.F. and Spronk, N. (eds.), Pergamon Press, Oxford, 103–114.

Butler, P.J. (1984) Physiological responses to voluntary diving: a bird's eye view. In *Proc. Icedive '84*, 77–89.

Butler, P.J. and Jones, D.R. (1982) The comparative physiology of diving in vertebrates. In *Advances in Comparative Physiology and Biochemistry*, Vol. 8, Lowenstein, O.E. (ed.), Academic Press, New York, 179–364.

Butler, P.J., West, N.H. and Jones, D.R. (1977) Respiratory and cardiovascular responses of the pigeon to sustained, level flight in a wind tunnel. *J. exp. Biol.* **71**, 7–26.

Butler, P.J. and Woakes, A.J. (1980) Heart rate, respiratory frequency and wing beat frequency of free flying barnacle geese, *Branta leucopsis*. *J. exp. Biol.* **85**, 213–226.

Butler, P.J. and Woakes, A.J. (1984) Heart rate and aerobic metabolism in Humboldt penguins, *Spheniscus humboldti*, during voluntary dives. *J. exp. Biol.* **108**, 419–428.

Cabot, J.B. and Cohen, D.H. (1980). Neural control of the avian heart. In *Hearts and Heart-like Organs*, vol. I, (ed.), Bourne, G.H., Academic Press, New York, 199–258.

Cade, T.J. and Greenwald, L. (1966) Nasal salt secretion in falconiform birds. *The Condor* **68**, 338–350.

Calder, W.A. and King, J.R. (1974) Thermal and caloric relations of birds. In *Avian Biology*, vol. IV, Farner, D.S. and King, J.R. (eds.), Academic Press, New York, 259–413.

Caputa, M. (1984) Some differences in mammalian versus avian temperature regulation: putative thermal adjustments to flight in birds. In *Thermal Physiology*, Hales, J.R.S. (ed.), Raven Press, New York, 413–417.

Cavanaugh, K.P. and Holmes, W.N. (1982) Effects of ingested petroleum on plasma levels of ovarian steroid hormones in photostimulated ducks. *Arch. Environ. Contam. Toxicol.* **11**, 503.

Card, L.E. and Nesheim, M.C. (1972) *Poultry Production*. Lea and Febiger, Philadelphia.

Chadwick, A. and Hall, T.R. (1983). Mechanism regulating secretion of prolactin in birds. In *Progress in Non-Mammalian Brain Research*, Nisticó, G. and Bolis, L. (eds.), CRC Press, Boca Raton, 80–95.

Chandola, A., Bhatt, D. and Pathak, V.K. (1983) Environmental manipulation of seasonal reproduction in Spotted munia (*Lonchura punctulata*). In *Avian Endocrinology: Environmental and Ecological Perspectives*, Mikami, S., Homma, K., and Wada, M. (eds.), Japan Scientific Societies Press, Tokyo, 229–242.

Cheng, M.F. and Silver, R. (1975) Estrogen-progresterone regulation of nest-building and incubation behaviour in ovariectomized ring doves (*Streptopelia risoria*). *J. Comp. Physiol. Psychol.* **88**, 256–263.

Crocker, A.D. and Holmes, W.N. (1971) Intestinal absorption in the duck (*Anas platyrhynchos*) maintained on freshwater and hypertonic saline. *Comp. Biochem. Physiol.* **40A**, 203–211.

Davies, S.J.J.F. (1979) The breeding season of birds in south-western Australia. *J. Roy. Soc. W. Aust.* **62**, 53–64.

Davies, S.J.J.F. (1982) Behavioural adaptations of birds to environments where evaporation is high and water is in short supply. *Comp. Biochem. Physiol.* **71A**, 557–566.

Dawson, W.R. (1982) Evaporative losses of water by birds. *Comp. Biochem. Physiol.* **71A**, 495–509.

Dawson, W.R. and Carey, C. (1976) Seasonal acclimatization to temperature in cardueline finches. I. Insulative and metabolic adjustments. *J. comp. Physiol.* **112**, 317–333.

Dawson, W.R. and Hudson, J.W. (1970) Birds. In *Comparative Physiology of Thermoregulation*, Vol. I, Whittow, G.C. (ed.), Academic Press, New York, 223–310.

Dawson, W.R., Marsh, R.L. and Yacoe, M.E. (1983) Metabolic adjustments of small passerine birds for migration and cold. *Am. J. Physiol.* **245**, R755–R767.

DeVoogd, T. and Nottebohm, F.W. (1981) Gonadal hormones induce dendritic growth in the adult avian brain. *Science* **214**, 202–204.

Dittami, J.P. (1981) Seasonal changes in the behavior and plasma titles of various hormones in barheaded geese *Anser indicus*. *Z. Tierpsychol.* **55**, 289–324.

Drent, R.H. (1975) Incubation. In *Avian Biology*, Vol. V, Farner, D.S. and King, J.R. (eds.), Academic Press, New York, 333–420.

Drent, R.H. and Daan, S. (1980) The prudent parents: energetic adjustments in avian breeding. *Ardea* **68**, 225–252.

El Halawani, N.E., Burke, W.H. and Dennison, P.T. (1980) Effect of nest-deprivation on serum prolactin level in nesting female turkeys. *Biol Reprod.* **23**, 118–123.

El Halawani, M.E., Burke, W.H., Dennison, P.T. and Silsby, J.L. (1982) Neuropharmacological aspects of neural regulation of avian endocrine function. In *Aspects of Avian Endocrinology: Practical and Theoretical Implications*, Scanes, C.G., *et al.* (eds.), Grad. Studies, Texas Tech. Univ. **26**, 33–40.

El-Soudi, K. and Al Jebouri, M.A.J. (1979) Production potential in subtropic climate of native Iraqi chickens compared to White Leghorn, New Hampshire and their cross, *Wld's Poult. Sci. J.* **35**, 227–235.

Emlen, S.T. (1975) Migration: orientation and navigation. In *Avian Biology*, Vol. V, Farner, D.S. and King, J.R. (eds.), Academic Press, New York, 129–219.

Ensminger, M.E. (1980) *Poultry Science*. Interstate, Danville, Illinois.

Ensor, D.M. (1978) *Comparative Endocrinology of Prolactin*. Chapman and Hall, London.

Etches, R.J. and Cheng, K.W. (1981) Changes in the plasma concentrations of luteinizing hormone, progesterone, oestradiol and testosterone and in the binding of follicle-stimulating hormone to the theca of follicles during the ovulation cycle of the hen (*Gallus domesticus*). *J. Endocrinol.* **91**, 11–12.

Ernst, S.A. and Ellis, R.A. (1969) The development of surface specialization in the secretory epithelium of the avian salt gland in response to osmotic stress. *J. Cell Biol.* **40**, 305–321.

Fänge, R., Schmidt-Nielsen, K. and Osaki, H. (1958) The nasal salt gland of the herring gull. *Biol. Bull. Mar. Biol. Lab. Woods Hole* **115**, 162–171.

Farner, D.S. (1964) The photoperiodic control of reproductive cycles in birds. *Am. Sci.* **52**, 137–156.

Farner, D.S. and Lewis, R.A. (1971) Photoperiodism and reproductive cycles in birds. *Photophysiology* **6**, 325–370.

Farner, D.S., Donham, R.S., Matt, K.S., Mattocks, P.W. Jr, Moore, M.C. and Wingfield, J.C. (1983) The nature of photorefractoriness. In *Avian Endocrinology: Environmental and Ecological Perspectives*, Mikami, S.I., Homma, K. and Wada, M. (eds.), Japan Scientific Societies Press, Tokyo, 149–166.

Fitzsimons, J.T. (1979) *The Physiology of Thirst and Sodium Appetite*. Monographs of the Physiological Society, No. 35, Cambridge University Press.

Fogden, M.P.L. (1972) The seasonality and population dynamics of equatorial forest birds in Sarawak. *Ibis* **111**, 307–343.

Ford, R.G., Wiens, J.A., Heinemann, D.L. and Hunt, G.L. (1982) Modelling the sensitivity to colonially breeding marine birds to oil spills: Guillemot and kittiwake populations on the Pribilof Islands, Bering Sea. *J. appl. Ecol.* **19**, 1–13.

Follett, B.K. (1973) Circadian rhythms and photoperiodic time-measurement. *J. Reprod. Fert. Suppl.* **19**, 5–18.

Follett, B.K. (1973) Photoperiodism and seasonal breeding in birds and mammals. In *Control of ovulation*, Crighton, D.B., Haynes, B., Foxcroft, G. and Lamming, G. (eds.), Butterworth, London, 267–293.

Follett, B.K. (1984) Birds, In *Marshall's Physiology of Reproduction* (3rd eds.), Lamming, G. (ed.), Churchill Livingstone, Edinburgh, 283–350.

Follett, B.K., Davies, D.T. and Gledhill, B. (1977) Photoperiodic control of reproduction in Japanese quail: changes in gonadotrophin secretion on the first day of induction and their pharmacological blockade. *J. Endocr.* **74**, 449–460.

Follett, B.K., Mattocks, P.W., Farner, D.S. (1974) Circadian function in the photoperiodic induction of gonadotrophin secretion in the white-crowned sparrow. *Proc. Natl. Acad. Sci. U.S.A.* **71**, 1666–1669.

Follett, B.K., Robinson, J.E., Simpson, S.M. and Harlow, C.R. (1981) Photoperiodic time measurement and gonadotrophin secretion in quail. In *Biological Clocks in Seasonal Reproductive Cycles*, Follett, B.K. and Follett, D.E. (eds.), Scientechnica, Bristol, 185–202.

Fraps, R.M. (1961) Ovulation in the domestic fowl. In *Control of Ovulation*, Villee, C.A. (eds.), Pergamon Press, New York, 133–162.

Fraser, H.M. and Sharp, P.J. (1975) Prevention of positive feedback in the hen (*Gallus domesticus*). *J. Endocrinol.* **76**, 181–182.

Gaunt, S.L.L. (1980) Thermoregulation in doves (Columbidae): a novel esophageal heat exchanger. *Science* **210**, 445–447.

Gilbert, A.B. (1979) Female genital organs. In *Form and Function in Birds*, King, A.S., and McLelland, J. (eds.), Academic Press, London, 237–360.

Goldsmith, A. (1983) Prolactin in avian reproductive cycles. In *Hormones and Behaviour in Higher Vertebrates*, Balthazart, J., Prove, E. and Gilles, R. (eds.), Springer, Berlin, pp.375–387.

Goldsmith, A. and Follett, B.K. (1980) Anterior pituitary hormones. In *Avian Endocrinology*, Epple, A. and Stetson, M.H. (eds.), Academic Press, London, 147–165.

Goldsmith, A.R. and Nicholls, T.J. (1984) Thyroxine induces photorefractoriness and stimulates prolactin secretion in European starlings (*Sturnis vulgaris*). *J. Endocrinol.* **101**, R1–R3.

Goldspink, G. (1977) Mechanics and energetics of muscle in animals of different sizes with particular reference to the muscle fibre composition of vertebrate muscle. In *Scale Effects in Animal Locomotion*, Pedley, T.J. (ed.) Academic Press, London, 37–55.

Griffin, D.R. (1969) The physiology and geophysics of bird navigation. *Q. Rev. Biol.* **44**, 255–276.

Gurney, M.E. and Konishi, M. (1980) Hormone-induced sexual differentiation of brain and behaviour in Zebra finches. *Science* **208**, 1380–1383.

Gwinner, E. (1981) Circannual systems. In *Handbook of Behavioural Neurobiology*, Vol. 4, *Biological Rhythms*, Aschoff, J. (ed.), Plenum, London and New York, 391–410.

Haase, E. and Donham, R.S. (1980) Hormones and domestication. In *Avian Endocrinology*, Epple, A. and Stetson, M.H. (eds.), Academic Press, New York and London, 549–565.

Haase, E., Sharp, P.J. and Paulke, E. (1982) The effects of castration on the seasonal pattern of plasma LH concentrations in wild mallard drakes. *Gen. Comp. Endocrinol.* **46**, 113–115.

Hamner, W.M. (1968) The photorefractory period of the house finch. *Ecology.* **49**, 211–227.

Hamner, W.M. (1971) On seeking an alternative to the endogenous reproductive rhythm hypothesis in birds. In *Biochronometry*, Menaker, M. (ed.), National Academy of Sciences, Washington D.C., 448–462.

Harding, C.F. and Follett, B.K. (1979) Hormone changes triggered by aggression in a national population of blackbirds. *Science* **203**, 918–920.

Hart, J.S. (1962) Seasonal acclimatization in four species of small wild birds. *Physiol. Zool.* **35**, 224–236.

Harvey, S. and Phillips, J.G. (1982a) Adrenocortical responses of ducks to treadmill exercise. *J. Endocr.* **94**, 141–146.

Harvey, S. and Phillips, J.G. (1982b) Endocrinology of salt gland function. *Comp. Biochem. Physiol.* **71A**, 537–546.

Herre, W. and Röhrs, M. (1983) Abstammung und Entwicklung des Hausflugels. In *Handbuch der Geflügelphysiologie*, Teil 1, Mehner, A. and Hartfield, W. (eds.), Karger, Basel and London, 19–53.

Hinde, R.A. and Steele, E. (1978) The influence of daylength and male vocalizations on the oestrogen-dependent behaviour of female canaries and budgerigars. *Adv. Study. Behav.* **8**, 40–74.

Holmes, W.N. (1972) Regulation of electrolyte balance in marine birds with special reference to the role of the pituitary adrenal axis in the duck (*Anas platyrhynchos*). *Fed. Proc.* **31**, 1587–1597.

Holmes, W.N. (1984) Petroleum products in the marine environment and their possible effects on sea birds. *Reviews in Environmental Toxicology*, 1, Hodgson, E. (ed.), Elsevier, Oxford, 251–317.

Holmes, W.N., Gorsline, J. and Cavanaugh, K.P. (1981) Some effects of environmental pollutants. In *Recent Advances in Avian Endocrinology (Advances in Physiological Sciences Vol. 33)* Pethes, G. Peczely, P. and Rudas, P. (eds.), Pergamon, Oxford, 1.

Holmes, W.N. and Phillips, J.G. (1985) The avian salt gland. *Biological Revs.* (in press).

Holmes, W.N. and Stewart, D.J. (1968) Changes in nucleic acids and protein composition of the nasal glands from the duck (*Anas platyrhynchos*) during the period of adaptation to hypertonic saline. *J. exp. Biol.* **48**, 509–519.

Hudson, D.M. and Bernstein, M.H. (1983) Gas exchange and energy cost of flight in the whitenecked raven, *Corvus cryptoleucus. J. exp. Biol.* **103**, 121–130.

Hughes, B.O. (1975) The concept of an optimum stocking density and its selection for egg production. In *Economic Factors Affecting Egg Production*, Freeman, B.M. and Boorman, K.N., (eds.), British Poultry Science, Edinburgh, 271–298.

Hutchison, R.E. (1975) Effects of ovarian steroids and prolactin on the sequential development of nesting behaviour in female budgerigars. *J. Endocr.* **67**, 29–39.

Hutt, F.B. (1949) *Genetics of the Fowl*. McGraw-Hill, New York and London.

Immelman, K. (1971) Ecological aspects of periodic reproduction. In *Avian Biology*, Vol. 1, Farner, D.S. and King, J.R. (eds.), Academic Press, London, 341–389.

Imai, K. (1983) Characteristics of rapid growth of the ovarian follicles in the chicken. In *Avian Endocrinology: Environmental and Ecological Perspectives*, Mikami, S., Homma, K., and Wada, M. (eds.), Japan Scientific Societies Press, Tokyo, Springer, Berlin, 39–65.

Inomoto, T. and Simon, E. (1981) Extracerebral deep-body cold sensitivity in the Pekin duck. *Am. J. Physiol.* **241**, R136–R145.

Ishii, S. (1980) Hormone-receptor interactions I: Peptide hormones. In *Avian Endocrinology*, Epple, A. and Stetson, M.H. (eds.), Academic Press, London, 1–15.

Johansen, K. and Millard, R.W. (1973) Vascular responses to temperature in the foot of the giant fulmar, *Macronectes giganteus*. *J. comp. Physiol.* **85**, 47–65.

Johansen, K. and Millard, R.W. (1974) Cold induced neurogenic vasodilatation in skin of the giant fulmar, *Macronectes giganteus*. *Am. J. Physiol.* **227**, 1232–1235.

Jones, J.H., Grubb, B. and Schmidt-Nielsen, K. (1983) Panting in the emu causes arterial hypoxaemia. *Resp. Physiol.* **54**, 189–195.

Jones, P.J. and Ward, P. (1976) The level of reserve protein as the proximate factor controlling the timing of breeding and clutch-size in the red-billed quelea, *Quelea quelea*. *Ibis* **118**, 547–574.

Keeton, W.T. (1981) The orientation and navigation in birds. In *Animal Migration*, Aidley, D.J. (ed.), Cambridge University Press, 81–104.

King, J.A. and Millar, R.P. (1982) Structure of avian hypothalamic gonadotrophin-releasing hormone. *S. Afr. J. Sci.* **78**, 124–125.

Klandorf, H. Stokkan, K.A. and Sharp, P.J. (1982) Plasma thyroxine and triiodothyronine levels during the development of photorefractoriness in willow ptarmigan (*Lagopus lagopus lagopus*) exposed to different photoperiods. *Gen. Comp. Endoc.* **46**, 281–287.

Kobayashi, H. and Takei, Y. (1982) Mechanisms for induction of drinking with special reference to angiotensin. II. *Comp. Biochem. Physiol.* **71A**, 485–494.

Kramer, G. (1957) Experiments on bird orientation and their interpretation. *Ibis* **99**, 196–227.

Kreithen, M.L. (1978) Sensory mechanisms for animal orientation—can any new ones be discovered? In *Animal Migration, Navigation and Homing* (ed. K. Schmidt-Koenig, K., and Keeton W.T.), (eds.), Berlin: Springer, Berlin, 25–34.

Lake, P.E. (1975) Gamete production and the fertile period with particular reference to domestic birds. *Symp. Zool. Soc. Lond.* **35**, 225–244.

Lake, P.E. (1981) Male genital organs. In *Form and Function in Birds*, King, A.S. and McLelland, J. (eds.), Academic Press, London, 1–61.

Lasiewski, R.C. (1963) Oxygen consumption of torpid, resting, active, and flying hummingbirds. *Physiol. Zool.* **36**, 122–140.

Lasiewski, R.C. and Dawson, W.R. (1967) A re-examination of the relation between standard metabolic rate and body weight in birds. *Condor* **69**, 13–23.

Le Maho, Y. (1977) The Emperor Penguin: a strategy to live and breed in the cold. *Am. Sci.* **65**, 680–693.

Lea, R.W., Dods, A.S.M., Sharp, P.J. and Chadwick, A. (1981) The possible role of prolactin in the regulation of nesting behaviour and the secretion of luteinizing hormone in broody bantams. *J. Endocrinol.* **91**, 89–97.

Lea, R.W., Sharp, P.J. and Chadwick, A. (1982) Daily variations in the concentrations of plasma prolactin in broody bantams. *Gen. Comp. Endocr.* **48**, 275–284.

Lehrman, D.S. (1965) Interaction between internal and external environments in the regulations of the reproductive cycle of the king dove. In *Sex and Behaviour*, Beach, F.A. (ed.), John Wiley & Sons, New York, 355–380.

Lehrman, D.S., Brody, P.N. and Wortis, R.P. (1961) The presence of the mate and of nesting material as stimuli for the development of incubation behaviour and for gonadotrophin secretion in the ring dove (*Streptopelia risoria*). *Endocrinology* **68**, 507–516.

Lin, M.T. and Simon, E. (1982) Properties of high Q_{10} units in the conscious duck's

hypothalamus responsive to changes in core temperature. *J. Physiol.* **322**, 127–137.
Lincoln, G.A., Racey, P.A., Sharp P.J. and Klandorf, H. (1980) Endocrine changes associated with spring and autumn sexuality of the rook (*Corvus frugilegus*). *J. Zool. Lond.* **191**: 137–153.
Lofts, B. (1964) Evidence of an autonomous reproductive rhythm in an equatorial bird (*Quelea quelea*), *Nature* **201**, 523–524.
Lofts, B. and Murton, R.K. (1968) Photoperiodic and physiological adaptation regulating avian breeding cycles and their ecological significance. *J. Zool. Lond.* **155**, 327–394.
Marey, E.J. (1890) *Le Vol des Oiseaux.* Masson, Paris.
Marrone, B.L. and Hertelendy, F. (1983) Steroidogenesis by avian ovarian cells: effects of luteinizing hormone and substrate availability. *Am. J. Physiol.* **244**, E487–E493.
Marder, J. (1983) Cutaneous evaporation—II. Survival of birds under extreme thermal stress. *Comp. Biochem. Physiol.* **75A**, 433–439.
Martin, G.R. and Young, S.R. (1984) The eye of the Humboldt penguin, *Spheniscus humboldti*: visual fields and schematic optics. *Proc. R. Soc. Lond. B.* **223**, 197–222.
Marshall, A., Hyatt, A.D., Phillips, J.G. and Condron, R.J. (1985) Isosmotic secretion and hyposmotic reabsorption in the avian nasal salt gland. *J. Memb. Physiol* (in press).
Matthews, G.V.T. (1968) *Bird Navigation* (2nd ed.), Cambridge University Press.
Mattocks, P.W., Farner, D.S. and Follett, B.K. (1976) The annual cycle of LH in the plasma of intact and castrated white-crowned sparrows. *Gen. Comp. Endocr.* **30**, 156–161.
Maung, Z.W. and Follett, B.K. (1977) Effect of chicken and ovine luteinizing hormone on androgen release and cyclic AMP production by isolated cells from the quail testis. *Gen. Comp. Endocr.* **33**, 242–253.
Meier, A.H. and Ferrell, B.R. (1978) Avian endocrinology in *Chemical Zoology*, Vol. 10, Florkin, M., Scheer, B., and Brush, J. (eds.), Academic Press, New York and London, 213–271.
Menaker, M., Hudson, D.J. and Takahashi, J.S. (1981) Neural and endocrine, components of circadian clocks in birds. In *Biological Clocks in Seasonal Reproductive Cycles*, Follett, B.K. and Follett, D.E. (eds.), Wright, Bristol, 171–183.
Merkel, von F.W. and Wiltschko, W. (1965) Magnetismus und Richtungsfinden zugunruhiger Rotkehlchen. *Vogelwarte* **23**, 71–77.
Meyer, C. (1974) Effects of lesions in the medial preoptic region on precocial copulation in the chick. *Horm. Behav.* **5**, 377–381.
Midtgård, U. (1983) Scaling of the brain and the eye cooling system in birds: a morphometric analysis of the *rete ophthalmicum*. *J. exp. Zool.* **225**, 197–207.
Morley, A. (1943) Sexual behaviour in birds from October to January. *Ibis.* **85**, 132–158.
Miyamoto, K., Hasegawa, Y., Nomura, M., Igarashi, M., Kangaura, K. and Matsuo, H. (1984) Identification of the second gonadotropin releasing hormone in chicken hypothalamus: evidence that gonadotropin secretion is probably controlled by two distinct gonadotropin-releasing hormones in avian species. *Proc Natl. Acad. Sci. U.S.A.* **81**, 3874–3878.
Murrish, D.E. (1973) Respiratory heat and water exchange in penguins. *Resp. Physiol.* **19**, 262–270.
Murrish, D.E. and Guard, C.L. (1977) Cardiovascular adaptations in the giant petrel, *Maronectes giganteus*, to the Antarctic environment. In *Adaptations within Antarctic Ecosystems*, Ilano, G.A. (ed.), Smithsonian Institution, Washington D.C., 551–530.
Murton, R.K. and Westwood, N.J. (1977) *Avian Breeding Cycles.* Clarenden Press, Oxford.
Newton, I. and Haas, M.B. (1984) The return of the Sparrowhawk. *Brit. Birds*, 47–70.
Nishida, T. (1980) Jungle fowl in south-east Asia. In *Biological Rhythms in Birds*, Tanable, Y., Tanaka, K. and Ookawa, T. (eds.), Japan Scientific Societies Press, Tokyo, 301–313.
Nottebohm, F. (1981) A brain for all seasons: cyclical anatomical changes in song control nuclei of the canary brain. *Science* **214**, 1368–1370.
Oliphant, L.W. (1983) First observations of brown fat in birds. *Condor* **85**, 350–354.
Opel, H. and Proudman, J.A. (1980) Failure of mammalian prolactin to induce incubation behavior in chickens and turkeys. *Poult. Sci.* **59**, 2550–2558.

Papi, F. (1982) Olfaction and homing in pigeons: ten years of experiments. In *Avian Migration* (ed.) Papi, F.H.G. Wallraff, H.G. (eds.), Springer Berlin, 149–159.
Papi, F., Fiore, L., Fiaschi, V. and Bevenuti, S. (1972) Olfaction and homing in pigeons. *Mon. Zool. Ital.* **6**, 85–95.
Pasquis, P., Lacaisse, A. and Dejours, P. (1970) Maximal oxygen uptake in four species of small mammal. *Resp. Physiol.* **9**, 298–309.
Paton, J.A. and Nottebohm, F.W. (1984) Neurons generated in the adult brain are recruited into functional circuits. *Science* **225**, 1040–1048.
Pavgi, S. and Chandola, A. (1981) Role of gonadal feedback in annual reproduction of the weaver bird: interaction with photoperiod. *Gen. Comp. Endocr.* **45**, 521–526.
Payne, R.B. (1972) Mechanisms and control of moult. In *Avian Biology*, Vol. 2, Farner, D.S. and King, J.R. (eds.), Academic Press, New York and London, 103–155.
Peaker, M. and Linzell, J.L. (1975) Salt glands in birds and reptiles. *Monographs of the Physiological Society*, No. 32, Cambridge University Press.
Pennycuick, C.J. (1969) The mechanics of bird migration. *Ibis* **111**, 525–556.
Pennycuick, C.J. (1975) Mechanics of flight. In *Avian Biology*, Vol. V, Farner, D.S. and King, J.R. (eds.), Academic Press, New York, 1–75.
Perrins, C.M. and Birkhead, T.R. (1983) *Avian Ecology*. Blackie, Glasgow and London.
Pinshow, B., Fedak, M.A., Battles, D.R. and Schmidt-Nielsen, K. (1976) Energy expenditure for thermoregulation and locomotion in emperor penguins. *Am. J. Physiol.* **231**, 903–912.
Prange, H.D. and Schmidt-Nielsen, K. (1970) The metabolic cost of swimming in ducks. *J. exp. Biol.* **53**, 763–777.
Prinzinger, R. and Hanssler, I. (1980) Metabolism-weight relationship in some small non-passerine birds. *Experientia* **37**, 1299–1300.
Ramirez, J.M. and Bernstein, M.H. (1976) Compound ventilation during thermal panting in pigeons: a possible mechanism for minimizing hypocapnic alkalosis. *Fed. Proc.* **35**, 2562–2565.
Rautenberg, W., May, B., Necker, R. and Rosner, G. (1978) Control of panting by thermosensitive spinal neurons in birds. In *Respiratory Function in Birds, Adult and Embryonic*, Piiper, J. (eds.), Springer Berlin, 204–210.
Rayner, J.M.V. (1979) A new approach to animal flight mechanics. *J. exp. Biol.* **80**, 17–54.
Rayner, J.M.V. (1980) Vorticity and animal flight. In *Aspects of Animal Movement*, Elder, H.Y. and Trueman, E.R. (eds.), Cambridge University Press, 177–199.
Richards, S.A. and Avery, P. (1978) Central nervous mechanisms regulating thermal panting. In *Respiratory Function in Birds, Adult and Embryonic*, Piiper, J. (ed.), Springer Berlin, 196–203.
Robinson, J.E. and Follett, B.K. (1982) Photoperiodism in Japanese quail. The termination of seasonal breeding by photorefractoriness. *Proc. Roy. Soc. Lond. B.* **215**, 95–116.
Rowan, W. (1931) *The Riddle of Migration*. Williams and Wilkins, Baltimore.
Saarela, S. and Vakkuri, O. (1982) Photoperiod-induced changes in temperature-metabolism curve, shivering threshold and body temperature in the pigeon. *Experientia* **38**, 373–374.
Saleyev, P. (1975) Ways of increasing goose meat production in the USSR. *Wld's Poult. Sci. J.* **31**, 276–287.
Saunders, D.S. (1977) *An Introduction to Biological Rhythms*. Blackie, Glasgow and London.
Scanes, C.G., Cheeseman, P., Phillips, J.G. and Follett, B.K. (1974) Seasonal and age variations of circulating immunoreactive luteinizing hormone in captive Herring gulls *Larus argentatus*. *J. Zool., Lond.* **174**, 369–375.
Scanes, C.G., Rabii, J. and Buonomo, F.C. (1982) Brain amines and the regulation of anterior pituitary secretion in the domestic fowl. In *Aspects of Avian Endocrinology: Practical and Theoretical Implications*. Scanes, C.G. *et al.* (eds), Grad. Studies, Texas Tech. Univ. **26**, 13–31.
Scheid, P. (1979) Mechanisms of gas exchange in bird lungs. *Rev. Physiol. Biochem. Pharmacol.* **86**, 137–184.

Scheid, P. (1982) Respiration and control of breathing. In *Avian Biology*, Vol. VI, Farner, D.S., King, J.R. and Parkes, K.C., (eds.), Academic Press, New York, 405–453.
Scheid, P. and Piiper, J. (1972) Cross-current gas exchange in avian lungs: effects of reversed parabronchial air flow in ducks. *Respir. Physiol.* **16**, 304–312.
Schmidt-Nielsen, K. (1960) Salt secreting gland of marine birds, *Circulation* **21**, 955–967.
Schmidt-Nielsen, K. (1972) Locomotion: energy cost of swimming, flying and running. *Science* **177**, 222–228.
Schmidt-Nielsen, K. (1983) *Animal Physiology: Adaptation and Environment* (3rd edn.). Cambridge University Press.
Schorger, N. (1966) *The Wild Turkey. Its History and Domestication.* University of Oklahoma Press, Oklahoma.
Schuchman, K.-L. (1979) Metabolism of flying hummingbirds. *Ibis* **121**, 85–86.
Sharp, P.J. (1980) Female reproduction. In *Avian Endocrinology*, Epple, A. and Stetson, M.H. (eds.), Academic Press, New York and London, 435–454.
Sharp, P.J. (1983) *Hypothalamic control of gonadotrophin secretion in birds*. In *Recent Progress in Non-mammalian Brain Research*, Nistico, G. and Bolis, L. (eds.), CRC Press, Boca Raton, 124–164.
Sharp, P.J. (1984) Seasonal breeding and sexual maturation. In *Reproductive Biology of Poultry* (Poultry Science Symposium 17), Cunningham, F.J., Lake, P.E. and Hewitt, D. (eds.), Longman Harlow, 203–217.
Sharp, P.J. and Gow, C.B. (1983) Neuroendocrine control of reproduction in the cockerel. *Poult. Sci.* **62**, 1671–1675.
Sharp, P.J. and Klandorf, H. (1984) Environmental and physiological factors controlling thyroid function in Galliformes. In *Environment and Hormones*, Ishii, S., Follett, B.K. and Chandola, A. (eds.), Japan Scientific Societies Press, Tokyo, 175–188.
Sharp, P.J. and Moss, R. (1981) A comparison of the responses of captive Willow Ptarmigan (*Lagopus lagopus lagopus*) red grouse (*Lagopus lagopus scoticus*) and hybrids to increasing day length with observations on the modifying effects of nutrition and on red grouse. *Gen. Comp. Endoc.* **45**, 181–188.
Sibley, R.N. (1981) Strategies of digestion and defaecation. In *Physiological Ecology. An Evolutionary Approach to Resource Use*, Townsend, C., and Calow, P. (eds.), Blackwell, Oxford, 109–139.
Silver, R. (1978) The parental behaviour of Ring doves. *Am. Sci.* **66**, 209–215.
Simon, E. (1982) The osmoregulatory system of birds with salt glands. *Comp. Biochem. Physiol.* **71A**, 547–556.
Simon-Opperman, Ch., Simon, E., Deutsch, H. Möhring, J. and Schoun, J. (1980) Serum arginine-vasotocin (AVT) and apparent and central control of osmoregulation in conscious Pekin ducks. *Pflugers Arch.* **387**, 99–106.
Skadhauge, E. (1974) Renal concentrating ability in selected West Australian birds. *J. exp. Biol.* **61**, 269–276.
Skadhauge, E. (1981) *Osmoregulation in Birds.* Springer, Berlin.
Snapp, B.D., Heller, H.C. and Gospe, S.M. (1977) Hypothalamic sensitivity in the California quail (*Lophortyx californicus*). *J. Comp. Physiol.* **117**, 345–357.
Stokkan, K.-A. and Sharp, P.J. (1980) Seasonal changes in the concentrations of plasma luteinising hormone and testosterone in Willow Ptarmigan (*Lagopus lagopus lagopus*) with observations on the effects of permanent short days. *Gen. Comp. Endocr.* **40**, 109–115.
Stokkan K.-A. and Sharp, P.J. (1984) The development of photorefractoriness in castrated willow ptarmigan. *Gen. Comp. Endocr.* **54**, 402–408.
Storey, C.R. and Nicholls, T.J. (1976) Some effects of manipulation of daily photoperiod upon the rate of onset of a photorefractory state in canaries *Serinus canarius. Gen. Comp. Endocr.* **30**, 204–208.
Sossinka, R. (1980) Ovarian development in an opportunistic breeder, the Zebra finch *Poephila guttata castanotis. J. exp. Zool.* **211**, 225–230.

Taylor, C.R. (1977) Exercise and environmental heat loads: different mechanisms for solving different problems? In *International Review of Physiology* (Environmental Physiology II), Vol. 15, Robertshaw, D., (ed.), 119–146.
Thapliyal, J.P. and Chandola, A. (1972) Thyroid in wild finches. *Proc. Natl. Acad. Sci. India*, **42** (B) Part 1, 76–90.
Thomas, D.H. (1982) Salt and water excretion by birds: the lower intestine as an integrator of renal and intestinal excretion. *Comp. Biochem. Physiol.* **71A**, 527–536.
Thomas, D.H. (1983) In *Veterinary Nephrology*, Hall, L.W. (ed.), Heineman London, 71–89.
Thomas, D.H. (1984) Sandgrouse as models of avian adaptations to deserts. *S. Afr. Tydskr. Dierk.* **19**, 113–120.
Thomas, D.H. and Robin, A.P. (1977) Comparative studies of thermoregulatory and osmoregulatory behaviour and physiology of five species of sandgrouse (Aves: Pteroclidae) in Morocco. *J. Zool. Lond.* **183**, 229–249.
Thomas, D.H., Pinshow, B. and Allan Degen, A. (1984) Renal and lower intestinal contributions to the water economy of desert dwelling phasianid birds: comparison of free living and captive chuckars and sand partridges. *Physiol. Zool.* **57** (1), 128–136.
Torre-Bueno, J.R. (1976) Temperature regulation and heat dissipation during flight in birds. *J. exp. Biol.* **65**, 471–482.
Torre-Bueno, J.R. and Larochelle, J. (1978). The metabolic cost of flight in unrestrained birds. *J. exp. Biol.* **75**, 223–229.
Tucker, V.A. (1966) Oxygen consumption of a flying bird. *Science* **154**, 150–151.
Tucker, V.A. (1968) Respiratory exchange and evaporative water loss in the flying budgerigar. *J. exp. Biol.* **48**, 67–87.
Urbanski, H. and Follett, B.K. (1982) Photoperiodic modulation of gonadotrophin secretion in castrated Japanese quail. *J. Endocrinol.* **92**, 73–83.
Uemura, H., Kobayashi, H., Okawara, Y. and Yamaguchi, K. (1983) Neuropeptides and drinking in birds. In *Avian Endocrinology: Environmental and Ecological Perspectives*, Mikami, S., et al. (eds.), Japan Scientific Societies Press, Tokyo Springer, Berlin, 225–262.
von Frisch, K. (1967). *The Dance Language and Orientation of Bees.* Oxford University Press, London.
von Saalfield, E. (1936) Untersuchungen uber das Hacheln bei Tauben. *Z. vergl. Physiol.* **23**, 727–743.
Walcott, C. (1982) Is there evidence for a magnetic map in homing pigeons? In *Avian Navigation*, Papi, F. and Wallraff, G. Springer, Berlin, 99–106.
Wallraff, H.G. (1983) Relevance of atmospheric odours and geomagnetic field to pigeon navigation: What is the 'map' basis? *Comp. Biochem. Physiol.* **76A**, 634–663.
Ward, P. (1969) The annual cycle of the Yellow-vented bulbul *Pycnonotus goivier* in a humid equatorial environment. *J. Zool. Lond.* **157**, 25–45.
Weathers, W.W. (1979) Climatic adaptation in avian standard metabolic rate. *Oecologia* **42**, 81–89.
Weibel, E.R. and Taylor, C.R. (eds.) (1981) Design of the mammalian respiratory system. *Resp. Physiol.* **44**, 1–164.
Weins, J.A., Ford, R.G., Heinemann, D. and Fieber, C. (1982) Guillemot and kittiwake populations on the Pribilof Islands, Alaska. In *Environmental Assessment of the Atlantic Continental Shelf*, Bureau of Land Management, U.S. Department of Commerce, Washington, D.C., 1.
Wentworth, B.C., Proudman, J.A., Opel, H., Wineland, M.J. Zimmermann, N.G. and Lapp, A. (1983) Endocrine changes in the incubating and brooding turkey hen. *Biol. Reprod.* **29**, 87–92.
West, N.H., Langille, B.L. and Jones, D.R. (1981) Cardiovascular system. In *Form and Function in Birds*, Vol. II, King, A.S. and McLelland, J. (eds.), Academic Press, London, 235–339.
Wieselthier, A.S. and Van Tienhoven, A. (1972) The effect of thyroidectomy on testicular size

and on the photorefractory period in the starling, *Sturnus vulgaris. J. exp. Zool.* **179**, 331–338.
Williams, J.B. and Sharp, P.J. (1978) Control of the preovulatory surge of luteinising hormone in the hen (*Gallus domesticus*): the role of progesterone and androgens. *J. Endocr.* **77**, 57–65.
Wilson, S.C. and Cunningham, F.J. (1984) Endocrine control of the ovulation cycle. In *Reproductive Biology of Poultry* (Poultry Science Symposium 17), Cunningham, F.J., Lake, P.E. and Hewitt, D. (eds.), Longman, Harlow, 29–49.
Wingfield, J.C. (1980) Fine temporal adjustment of reproductive functions. In *Avian Endocrinology*, Epple, A. and Stetson, M.H. (eds.), Academic Press, London and New York, 369–389.
Wingfield, J.C. (1983) Environmental and endocrine control of avian reproduction: an ecological approach. In *Avian Endocrinology: Environmental and Ecological Perspectives*, Mikami, S., Homma, K. and Wada, M. (eds.), Japan Scientific Societies Press, Tokyo and Springer, Berlin, 265–288.
Withers, P.C. (1977) Respiration, metabolism and heat exchange of euthermic and torpid poorwills and hummingbirds. *Physiol. Zool.* **50**, 43–52.
Woakes, A.J. and Butler, P.J. (1983) Swimming and diving in tufted ducks, *Aythya fuligula*, with particular reference to heart rate and gas exchange. *J. exp. Biol.* **107**, 311–329.
Wolf, L.L. and Hainsworth, F.R. (1972) Environmental influence on regulated body temperature in torpid hummingbirds. *Comp. Biochem. Physiol.* **41A**, 167–173.
Wolford, J.H. (1984) Induced moulting in laying fowls. *Wld's Poult. Sci. J.* **40**, 66–73.
Wooller, R.D. and Dunlop, J.N. (1979) Multiple laying by the silver gull, *Larus novaehollandiae* Stephens on Carnac Island, Western Australia. *Aust. Wldl. Res.* **6**, 325–335.
Wright, A., Holmes, W.N. and Gorsline, J. (1982) Colostomy of the duck (*Anas platyrhynchos*) a surgical technique and an assessment of its chronic effects on osmotic balance. *Comp. Biochem. Physiol.* **72A**, 663–668.
Wyndham, E. (1980) Aspects of biorhythms in the Budgerigar *Melopsittacus undulatus* (Shaw) a parrot of inland Australia. In *XVII Internat. Cong. Ornithol.*, Nohring, R. (ed), Deutschen Ornithologen Gesellschaft, Berlin, 485–492.

Index

Abbreviations
    follicle-stimulating hormone   FSH
    gonadotrophin-releasing hormone GnRH
    luteinizing hormone   LH
    preoptic/anterior hypothalamus   POAH
    rete mirabile ophthalmicum   RMO
Albatross (*Diomedea exulans*)
    breeding cycles   143
    muscles, gliding   25–6
    temperature   79
American Goldfinch (*Carduelis* (*Spinus*) *tristis*)
    breeding   63–4
    shivering   149
Anseriformes (ducks, geese) *and see* under species
    domestication   179–81
    phallus   116
Apterygiformes (kiwis, *Apteryx* sp.)
    body temperature   57
Arctic Tern (*Sterna paradisaea*)
    migration   41

Bar-headed Goose (*Anser indicus*)
    migration   43
    oxygen and altitude   23–4
    sexual maturity   142
Barnacle Goose (*Branta leucopsis*)
    flight tests   25
    respiration   20, 22
    ventilation   22
Bateleur Eagle (*Terathopius ecaudatus*)
    salt gland   88
    bats cf. Aves   23
Bedouin Fowl (Galliformes — ? *Numida meleagris*)
    panting   74
Blackbird (*Turdus merula*)
    partial migrant   141
Black Brent Goose (*Branta bernicla*)
    migration   43
Black-capped Chicadees (Black-cap titmouse, *Parus atricapillus*)
    shivering   61–4
Black Duck (Black-headed duck, *Heteronetta atricapilla*)
    heat loss   76
    wingbeat/respiration   21
Black-headed Grosbeak (*Pheucticus melanocephalus*)
    water intake   84
Blackpoll Warbler (*Dendroica striata*)
    migration   43
    power energy   13
Blue Tit (*Parus caeruleus*)
    breeding   146–7
Bobolink (*Dolichonyx orizivorus*)
    breeding   149
Breeding
    autonomous rhythms   162–4
    circa-annual   143
    continuous   142–3
    delayed   142
    factors, proximate   151–5
    hormones   148–50, 152–7, 170–3
    non-annual   142–5
    opportunity   143–5
    photorefractoriness   146–50, 167–73
    scotorefractoriness   166–7
    seasonal   146–51, 165–6
    tropics   145–6, 164–5
brood parasitism   131
Brown Towhee (*Pipilo fuscus*)

water intake 84
Budgerigar (*Melopsittacus undulatus*)
   breeding 143
   incubation 137
   power input 13–14, 17
   shivering 62
   water intake 84

California Quail (*Callipepla californica*; *Lophortyx californica*)
   cold, brood patch 66
   water intake 84
Canada Goose (*Branta canadensis*)
   ventilation 22
Canary (*Serinus canarius*)
   and daylength 168–9, 170–3
   incubation 133–5
   sexual behaviour 124
Cardinal (*Cardinalis cardinalis*)
   critical temperature 60
Cassowary (*Casuarius casuarius*) see Casuariiformes
Casuariiformes (Emu and Cassowary, q.v.)
   body temperatures 57
Chaffinch (*Fringilla coelebs*)
   fat 51
   migration 53
Chiffchaff (*Phylloscopus collybita*)
   rhythms 165
Chukar (partridge, *Alectoris chukar*, see Phasianidae)
   and intestinal fluid 109
Circulatory system
   cardiovascularity 22–3
   Fick's formula 22–3
   heart 6–7
   heat loss 69–70, 72–3
   high altitude 23–4
   in Antarctica, cold 67–8
   nervous system 6–7
   receptors 7
   RMO (*see* Abbr.) 67, 71–2
   thermoregulation 64–6
cockerel (*see under* domestic fowl)
Columbidae (doves, pigeons, q.v.)
   breeding cycles 146
   heat loss 76
Common Murre (*Uria aalge*)
   territory 153
Coot (*Fulica atra*)
   respiration 66
Crane (*Grus grus*)
   migration 54

Crossbill (*Loxia curvirostra*)
   breeding 150

daylength
   circadian rhythms 159–62, 169–70
   critical 157–9
   general 152
   Zeitgeber 156, 160
desert phasianids (Phasianidae: e.g. Chukar, Sand Partridge)
   kidney 107–9
   water economy 109
Dipper (*Cinclus cinclus*)
   critical temperature 60
domestication
   broodiness 184–5
   chickens 174–7
   ducks and geese 179–81
   farming, intensive 181–3
   general 174–81
   induced moult 185
   performance 185–7
   turkeys 178–9
domestic fowl (*Gallus gallus*)
   breeding 147
   broodiness 184–5
   circadian rhythms 128, 130–1
   cursorial 27
   domestication 174–7
   GnRH (*see* Abbr.) 118
   incubation 136, 138
   intestine 103, 104
   kidney 108
   LH (*see* Abbr.) 120–2
   lighting 181–3
   male (cockerel) semen 116
   moult 185
   ovulation 127, 129–30
   prolactin 122–3
   reproduction 113–5
   respiration 77
   sexual behaviour 123–4
   stocking density 185–7

eagles (Accipitridae)
   breeding cycle 143
   thermal soaring 55
Eider (*Somateria mallissima*)
   salt gland 88–9
Eleonora's Falcon (*Falco eleonorae*)
   breeding 150
Emu (*Dromaius novae-hollandiae, and see* Casuariiformes)
   cursorial 27–8

INDEX

POAH (*see* Abbr.) 78
environment and reproduction
    breeding (q.v.) 142–73
    breeding, general strategies 140–1
    migration 141–2
    oil, contamination by 189–96
    pesticides 196–8
    pollution 187–9
Evening Grosbeak (*Coccothraustes vespertina*)
    critical temperature 60

Fish Crow (*Corvus ossifragus*)
    power input 14–15, 17, 19
    respiration 20–2
    ventilation 74
flight, economics of
    fat 51
    flapping 76–7
    formation 55
    fuel 51–2
    tail winds 53–4
    thermal air 54–5
flight energetics
    air movement 9
    flying behaviour 25–6
    high altitude 23–4
    kinetics 11–12
    metabolic rate 13–16
    migration 11–12
    models 9–10
    oxygen 13–14, 16–18, 19–20, 25–6
    physiology 18–22
    power input and output 9–16
    velocity 11–16
Fox Sparrow (*Passerella iliaca*)
    migration 41
    water intake 84

Garden Warbler (*Sylvia borin*)
    rhythms 165
Goat Suckers (Poorwill; *Phalaenoptilus nuttallii*)
    torpor 57–60
Golden Eagle (*Aquila chrysaetos*)
    semen 116
goose, domestic (*Anser domesticus* and see domestication)
    fertile period 115
Gray Catbird (*Dumetella carolinensis*)
    muscles 51
Greater Flamingo (*Phoenicopterus ruber*)
    ventilation 75
Greater Shearwater (*Puffinus gravis*)
    migration 53
grebes (Podicipedidae)
    body temperatures 57
Grey Teal (*Anas australis*)
    breeding 144
guinea fowls (Numididae)
    fertile period 115

hawks (Accipitridae)
    and pigeons 111
    sexual maturity 141
Herring Gull (*Larus argentatus*)
    gliding oxygen 25
    heat loss, dry 69–70
    muscle fibres 27
    salt gland 89–93
    sexual maturity 142
House Finch (*Carpodacus mexicanus*)
    water intake 84
House Sparrow (*Passer domesticus*)
    breeding 149
    day length 168–9
    muscle fibres 27
    water intake 84
hummingbirds (Trochilidae spp.; *Panterpe insignis, Eugenes fulgens, Amazilia cyaniformes, A. tzactl*)
    hovering 62
    oxygen 25
    torpor 57–60

Indigo Bunting (*Passerina cyanea*)
    migration 141
    navigation 46, 50

Japanese Quail (*Coturnix japonica*)
    breeding 147
    FSH (*see* Abbr.) 119
    LH (*see* Abbr.) 119
    ovulation 127
    photoperiodism 156–9
    photoreceptors 125–6
    POAH (*see* Abbr.) 83
    scotorefractoriness 166–7
    shivering 62
    substance P 83
    testosterone 119
juncos (Dark-eyed; Slate-coloured; *Junco hyemalis*)
    breeding 149
    migration 141
    photoperiodism 156–7
    scotorefractoriness 166–7
    water intake 84

Kestrel (*Falco sparverius*)
  brain temperature 71
  breeding 150
  power input 17
Kittiwake (*Rissa tridactyla*)
  respiration 66
Kiwi (*Apteryx australis*)
  ovary 113

Laughing Gull (*Larus atricilla*)
  flapping flight 25
  power input 17
  respiration 13
Lesser Snow Goose (*Anser caerulescens*)
  migration 43
locomotion
  circulatory system 6–8
  diving 33–40
  energetics 8–16
  flying, types 18–27
  land and water 27–40
  oxygen 16–18, 23–4, 25–6
  respiratory system 3–6
  skeletal muscles 8
  swimming 30–2
  walking/running 27–30, 67–8
loons (Gaviiformes, *Gavia immer*)
  body temperature 57

Mallard Duck (Pekin duck; *Anas platyrhynchos*)
  altitude 23–4
  breeding 146–8
  cold 64
  eggs 126
  fertile period 115
  intestinal absorption 87
  kidney 108
  locomotion (q.v.) 27
  navigation 46
  ovulation 126–7
  photorefractoriness 172–3
  POAH (*see* Abbr.) 78–9
  respiration 66
  rhythm 162–4
  salt gland 92–8
  scotorefractoriness 166–7
  territory 153
  ventilation 74, 77–8
migration
  altitude 43
  artificial induction 141–2
  behaviour and gonads 141–2
  biannual 141

  daylength 141–2, 155
  distances 41–3
  fat deposits 141, 165
  infra-sounds 48
  integration, clues 49–50
  magnetic fields 46–7
  navigation 44–50
  odours 48–9
  orientation 43–4
  patterns 41
  proximate factors 140–1
  routes 52–3
  tail winds 53–4
  visual clues 44–6
  Zugenruhe 141, 165
Mourning Dove (*Zenaida macroura*)
  water intake 84
Munia Finches (Estrildidae))
  breeding 150–1
  rhythms 165–6
  thyroid 125
  water intake 84
Mute Swan (*Cygnus olor*; *see* Swans)
  panting 74
  wingbeat 16

Noddy Tern (*Anous stolidus*; Sternidae)
  salt gland 102
Northern Fulmar (*Fulmarus glacialis*)
  fertile period 115

osmoregulation
  behaviour 110–11
  homeostasis 81–3, 85–9, 98–102
  intestine 102–5, 111
  kidney 102, 105–10
  nitrogen 109–10
  salinity 85
  salt glands 89–102
  thirst and salt 83–4
  water drinking 84–5
ostriches (Struthioniformes; *Struthio camelus*)
  body temperature 57
  incubation 131
  ventilation 75–6

Painted Quail (*Excalfactoria chinensis*)
  heat loss 73
Parrot, Galah (Roseate Cockatoo; *Kakatoe roseicapilla*)
  cloaca/rectum 104–5
  kidney 107
Partridge (*Perdix perdix*)

heat loss 76
scotorefractoriness 166−7
passerines (Passeriformes)
  breeding 145−6
  eggs 126
Pekin Duck (*see* Mallard)
penguins (Spheniscidae)
  body temperature 57
  in Antarctica 67−8
  locomotion 27−8, 32
  Adélie (*Pygoscelis adeliae*)
    diving 37, 39
    panting, gular flutter 74
    POAH (*see* Abbr.) 78
  Chinstrap (*Pygoscelis antarctica*)
    panting 74
  Emperor (*Aptenodytes forsteri*)
    breeding 150−1
    cold 60−1
    diving 37
  Gentoo (*Pygoscelis papua*)
    diving 39
    panting 74
  Humboldt (*Spheniscus humboldti*)
    diving 33, 35, 37−40
  King (*Aptenodytes patagonicus*)
    breeding 143
    energy and cold 62−3
  Rockhopper (*Eudyptes crestatus*)
    diving 39
  Royal (*Eudyptes chrysolopins schlegeli*)
    diving 39
petrels (Procellariidae)
  cold 64−6
  core temperature 79
  egg 126
  fertile period 115
  hovering 25
  muscles 25−6
Pheasant (*Phasianus colchicus*)
  wingbeat/respiration 21
Pied Flycatcher (*Ficedula hypoleuca*)
  navigation 50
  territory 153
Pied Kingfisher (*Ceryle rudis*)
  hovering 25
pigeons (Columbidae)
  and hawks 111
  brain temperature 71−2
  fat 13
  flight patterns 24−5
  heat loss 76
  interactions 155
  muscle fibres 26−7

navigation 44−50
oxygen 22
panting 73−5, 77−8
POAH (*see* Abbr.) 78−9
power output 9−12, 17
respiration 21−2
shivering 64
substance P 83
Pink-eared Duck (*Malacorhynchus membranaceus*)
  breeding 144
plovers (Charadriidae)
  incubation 131
procellids (Procellariiformes; albatrosses, shearwaters, petrels)
  body temperature 57
Purple Sandpiper (*Calidris maritima*)
  moult 138

Quails (*see also under* species of Phasianidae)
  fertile period 115
  POAH (*see* Abbr.) 77−8
  sexual behaviour 123−4
  wingbeat/respiration 21

raptors (=Falconiformes; hawks, eagles etc.)
  hovering 25
Ratites (=Struthionidae and Casuariidae; ostrich, cassowary)
  phallus 116
  sexual maturity 141
Red-billed Quelea (*Quelea quelea*)
  breeding 144−6
Red Grouse (*Lagopus lagopus*)
  food 153−4
Redpoll (*Carduelis flammea*)
  migration 41
Redstart (*Phoenicurus phoenicurus*)
  wingbeat 16
Red-winged Blackbird (*Agelaius phoeniceus*)
  food 153
  territory 153
reproduction (*see also* environment)
  behaviour 123−4
  brain 117−18
  chronobiology 126−30
  circadian rhythms 128
  gonads 113−17
  hormonal control 118−24, 129−30, 132−9
  incubation and brooding 131−8
  moult 138−9, 165, 185
  ovary 113−15
  oviduct 115−16

photoreceptors 125–6
testis 116–17
thyroid 124–5
respiratory system
  air flow 5
  blood system 3–6
  bronchi 3–4
  carotid body 5–7
  CO_2 5–6
  evaporation
    CO_2 74, 77–8
    panting 73–5, 77–8
    skin 73
    water 73
  flying 18–22
  lungs 3–5
  thermoregulation 66
  ventilation 74–6
Rhea (Common or Greater; *Rhea americana*)
  cursorial 27
  heat loss 76
Ring-billed Gull (*Larus novaehollandiae*)
  navigation 47
Ring Dove (*Streptopelia risoria*)
  fertile period 115
  incubation and brooding 132–3, 135–6, 138
  interactions 155
  sexual behaviour 123–4
Ring-necked Pheasant (*Phasianus colchicus*)
  fertile period 115
Roadrunner (*Geococcyx californianus*)
  heliothermia 60
  salt gland 88
Robin (*Erithacus rubecula*)
  Zugenruhe, in caged 47
Rook (*Corvus frugilegus*)
  breeding 148–9
  scotorefractoriness 166–7
Ruff (*Philomachus pugnax*)
  migration 52–3
Ruffed Grouse (*Bonasa umbellus*)
  shivering 61–4

salt glands (*see also under* various species)
  anatomy 89–91
  cells 91–3
  control 97–8
  hormonal role 98–102
  innervation 91
  secretory mechanisms 95–7
  stimulation 94–5
  vascularization 91
sand grouse (Pteroclididae)

  behaviour 110–11
Savannah Sparrows (*Passerculus sandwichensis* and subspp. *P. s. beldingi, rostratus, brooksi*)
  kidney 107
  rectum 103
  water intake 84
Scrub-fowl (Brush-Turkey; Megapodes)
  egg care 131
shearwaters (Procellariidae *and see under* species)
  eggs 126
Silver Gull (*and see* Ring-billed)
  breeding 150–1
Snow Bunting (*Plectrophenax nivalis*)
  critical temperature 60
song sparrows (*Melospiza* spp.)
  food (*M. melodia*) 153–4
  water intake (*M. cooperi; M. maxillaris; M. samuelis*) 84
Sooty Tern (*Sterna fuscata*)
  breeding 143
sparrows (*Zonotrichia atricapilla; Z. albicollis*)
  water intake 84
Sparrowhawk (*Accipiter nisus*)
  pesticides 198
  salt gland 82
Starling (*Sturnus vulgaris*)
  day length 168–9
  heat loss 76
  kidney 107
  migration 43–4
  power input 15, 17
  respiration 20
  scotorefractoriness 166–7
  thyroid 125
storks (Ciconiidae)
  locomotion 27
  thermal soaring 55
stress (*see under* flight, respiration etc.)
  cold 67–8
  heat 73–7
swallows and martins (Hirundinidae)
  breeding 149
  migration 41
swans (general; Anatidae)
  daylength 157–9

thermoregulation
  body temperature 56–7
  brain temperature 70–2
  cold exposure 60–1
  hypothalamus 78–9

incubation and cold 67
  nervous mechanisms 68–9
Thrasher (*Toxostoma dorsale*)
  water intake 84
Tufted Duck (*Aythya fuligula*)
  diving 33–8
  locomotion 27, 30–2
Turkey (*Meleagris gallopavo*)
  broodiness 184–5
  domestication 178–9
  fertile period 115
  incubation 133–6
  locomotion 28–30
  moult 185
  ovulation 127
  prolactin 122–3
  stocking density 185–7

vultures (Accipitridae; Falconiformes)
  breeding cycle 143
  heliothermia, in Turkey Vulture (*Cathartes avra*) 60

Waxbill sp. (*Estrilda troglodytes*)
  water intake 84
weaver finches (Ploceinae)
  photorefractoriness 172
Wheatear (*Oenanthe oenanthe*)
  migration 42–3
White-crowned Sparrow (*Zonotrichia leucophrys*)
  breeding 149, 151–3
  food 154
  migration 141–2
  photoreceptors 125–6
  territory 153
White-necked Raven (*Corvus cryptoleucus*; also chihuahuan raven)
  heat loss 76–7
  heat storage 70
  power input 17, 19
  ventilation 75
White-throated Sparrow (*Zonotrichia albicollis*)
  circadian rhythm 161–2, 169–73
  interactions 155
  migration 142
  navigation 46, 50
  photoperiodism 155–9
  photorefractoriness 167–8
Whooper Swan (*see* swans)
  migration 42, 43
Willow Ptarmigan (*Lagopus lagopus*)
  breeding 149, 152

daylength 168–73
shivering 64

Zebra Finch (*Taeniopygia guttata*)
  breeding 143–4
  heat loss 73
  kidney 107
  RMO (*see* Abbr.) and brain coolness 71
  sexual behaviour 124
  water intake (as *Poephila guttata*) 84
Zeitgeber 156, 160
Zoological nomenclature
  Accipiter nisus, Sparrow Hawk
  Agelaius phoeniceus, Red-winged Blackbird
  Alectoris chukar, Chukar
  Amazilia cyaniformes, Hummingbird
  Amazilia tzactl, Hummingbird
  Anas australis, Grey Teal
  Anas platyrhynchos, Mallard Duck
  Anous stolidus, Noddy Tern
  Anser caerulescens, Lesser Snow Goose
  Anser domesticus, Domestic Goose
  Anser indicus, Bar-headed Goose
  Aptenodytes forsteri, Emperor Penguin
  Aptenodytes patagonicus, King Penguin
  Apteryx australis, Kiwi
  Aquila chrysaetos, Golden Eagle
  Aythya fuligula, Tufted Duck
  Bonasa umbellus, Ruffed Grouse
  Branta bernicla, Black Brent Goose
  Branta canadensis, Canada Goose
  Branta leucopsis, Barnacle Goose
  Calidris maritima, Purple Sandpiper
  Callipepla californica, California Quail
  Cardinalis cardinalis, Cardinal
  Carduelis flammea, Redpoll
  Carduelis (*Spinus*) *tristis*, American Goldfinch
  Carpodacus mexicanus, House Finch
  Casuarius casuarius, Cassowary
  Cathartes avra, Turkey Vulture
  Ceryle rudis, Pied Kingfisher
  Cinclus cinclus, Dipper
  Coccothraustes vespertina, Evening Grosbeak
  Corvus cryptoleucus, White-necked Raven
  Corvus frugilegus, Rook

INDEX

Corvus ossifragus, Fish Crow
Coturnix japonica, Japanese Quail
Cygnus olor, Mute Swan
Dendroica striata, Blackpoll Warbler
Diomedea exulans, Albatross
Dolichonyx orizivorus, Bobolink
Dromaius novae-hollandiae, Emu
Dumetella carolinensis, Gray Catbird
Erithacus rubecula, Robin
Estrilda troglodytes, Waxbill
Eudyptes chrysolopins schlegeli, Royal Penguin
Eudyptes crestatus, Rockhopper Penguin
Eugenes fulgens, Hummingbird
Excalfactoria chinensis, Painted Quail
Falco eleonorae,, Eleonora's Falcon
Falco sparverius, Kestrel
Ficedula hypoleuca, Pied Flycatcher
Fringilla coelebs, Chaffinch
Fulica atra, Coot
Fulmarus glacialis, Northern Fulmar
Gallus gallus, domestic fowl
Gavia immer, Loon
Geococcyx californianus,, Roadrunner
Grus grus, Crane
Heteronetta atricapilla, Black Duck (Black-headed)
Junco hyemalis, Junco (dark-eyed; slate-coloured)
Kakatoe roseicapilla, Galah Parrot
Lagopus lagopus, Red Grouse, Willow Ptarmigan
Larus argentatus, Herring Gull
Larus atricilla, Laughing Gull
Larus novaehollandiae, Ring-billed Gull
Lophortyx californica, California Quail
Loxia curvirostra, Crossbill
Malacorynchus membranaceus, Pink-eared Duck
Meleagris gallopavo, Turkey
Melopsittacus undulatus, Budgerigar
Melospiza cooperi, Song Sparrow
Melospiza maxillaris, Song Sparrow
Melospiza melodia, Song Sparrow
Melospiza samuelis, Song Sparrow
Numida meleagris, ? Bedouin Fowl
Oenanthe oenanthe, Wheatear
Panterpe insignis, Hummingbird
Parus atricapillus, Black-capped Chicadee (Black-cap Titmouse)
Parus caeruleus, Blue Tit
Passer domesticus, House Sparrow
Passerculus sandwichensis, Savannah Sparrow
Passerculus sandwichensis beldingi, Savannah Sparrow
Passerculus sandwichensis brooksi, Savannah Sparrow
Passerculus sandwichensis rostratus, Savannah Sparrow
Passerella iliaca, Fox Sparrow
Passerina cyanea, Indigo Bunting
Perdix perdix, Partridge
Phalaenoptilus nuttallii, Goat Sucker
Phasianus colchicus, Ring-necked Pheasant
Pheucticus melanocephalus, Black-headed Grosbeak
Philomachus pugnax, Ruff
Phoenicopterus ruber, Greater Flamingo
Phoenicurus phoenicurus, Redstart
Phylloscopus collybita, Chiffchaff
Pipilo fuscus, Brown Towhee
Plectrophenax nivalis, Snow Bunting
Puffinus gravis, Greater Shearwater
Pygoscelis adeliae, Adélie Penguin
Pygoscelis antarctica, Chinstrap Penguin
Pygoscelis papua, Gentoo Penguin
Quelea quelea, Red-billed Quelea
Rhea americana, Rhea
Rissa tridactyla, Kittiwake
Serinus canarius, Canary
Somateria mallissima, Eider
Sphensicus humboldti, Humboldt Penguin
Sterna fuscata, Sooty Tern
Sterna paradisaea, Arctic Tern
Streptopelia risoria, Ring Dove
Struthio camelus, Ostrich
Sturnus vulgaris, Starling
Sylvia borin, Garden Warbler
Taeniopygia guttata, Zebra Finch
Terathopius ecudatus, Bateleur Eagle
Toxostoma dorsale, Thrasher
Turdus merula, Blackbird
Uria aalge, Common Murre
Zenaida macroura, Mourning Dove
Zonotrichia albicollis, White-throated Sparrow
Zonotrichia atricapilla, Sparrow
Zonotrichia leucophrys, White-crowned Sparrow